Cover Crops in
Smallholder Agriculture

Cover Crops in Smallholder Agriculture

Lessons from Latin America

Simon Anderson, Sabine Gündel and
Barry Pound with Bernard Triomphe

Practical Action Publishing Ltd
25 Albert Street, Rugby, CV21 2SD, Warwickshire, UK
www.practicalactionpublishing.org

© Intermediate Technology Publications 2001

First published 2001\Digitised 2013

ISBN 10: 1 85339 530 7
ISBN 13: 9781853395307
ISBN Library Ebook: 9781780442921
Book DOI: http://dx.doi.org/10.3362/9781780442921

A catalogue record for this book is available from the British Library.

The authors, contributors and/or editors have asserted their rights under
the Copyright Designs and Patents Act 1988 to be identified as authors of
their respective contributions.

Since 1974, Practical Action Publishing has published and disseminated
books and information in support of international development work
throughout the world. Practical Action Publishing is a trading name
of Practical Action Publishing Ltd (Company Reg. No. 1159018), the
wholly owned publishing company of Practical Action. Practical Action
Publishing trades only in support of its parent charity objectives and any
profits are covenanted back to Practical Action (Charity Reg. No. 247257,
Group VAT Registration No. 880 9924 76).

Contents

Acknowledgements

This book arose out of a regional workshop held in Yucatan, Mexico during February 1997 (Anderson et al., 1997). The participants, who were *campesino* producers, extensionists, researchers and non-governmental organization (NGO) workers, came from Mexico, Honduras, Guatemala, Nicaragua, El Salvador, Paraguay, Bolivia, the USA and Europe. They met to discuss the present and potential application of cover crops in small-scale farming systems, based on practical experience and research involvement. The organizing committee consisted of representatives from the Autonomous University of Yucatan, the Natural Resources Institute (NRI), Humboldt University, Berlin, and Wye College, London University. The workshop provided a forum where information, which addressed situations faced in rural communities, as well as methodological aspects of research and extension, was presented and analysed.

Financial support was provided by DFID and GTZ. Facilities and logistical support were provided from the FMVZ-UADY.

Our thanks go to all the people who contributed to the success of the meeting and to the further diffusion of the outcomes. Special thanks are due to all the workshop participants, who provided the broad experiences on which this book is based.

The publishers would like to thank Professor Jules Pretty for the use of the cover photograph.

Preface

The Merida Workshop

In February 1997 a four-day regional workshop was held in Merida, Yucatan, on cover crops as components of smallholder agriculture. The aims of the workshop were to:

- share experiences beween different actors involved in the use of cover crops
- systematize existing experiences on the integration of cover crops within agricultural systems under different conditions
- facilitate the integration of different perspectives, objectives and methods
- discuss the implications for research and development
- facilitate future collaboration between projects and institutions.

The workshop was structured around five main themes, which were introduced through the consideration of selected case studies presented by workshop participants (Appendix 1 contains a list of original case studies). The themes were:

- cover crops in annual cropping systems
- cover crops in perennial cropping systems
- cover crops for soil and water conservation
- cover crops and livestock husbandry
- diffusion and adaptation of cover crop innovations.

The workshop's 45 participants (see Appendix 2) – who were *campesino* producers, extensionists, researchers and NGO workers – came from Mexico, Honduras, Guatemala, Nicaragua, El Salvador, Paraguay, Bolivia, the USA and Europe. Apart from their experiences they brought a range of materials to the workshop, including posters, photos, videos and other visual materials, which were presented during an open day. This informal presentation was joined by representatives from local research institutions and farmers' groups, who were also invited to present posters on their activities.

The reason for this book

Lessons learned in Latin America about the use and dissemination of cover crops in different agroecosystems need to be made more widely available not only to Spanish-speaking, but also to Anglophone, regions. This publication aims to inform a wide range of actors involved in rural development projects, as well as those in applied research, of the potential of cover crops as components of smallholder agriculture systems. The objective of the book is to share recently acquired knowledge with practitioners, and to motivate others to discover, and to experiment with, new forms of cover crop use. There are several important initiatives being taken by different organizations to enhance the development and diffusion of knowledge on cover crops. However, the results and experiences of those initiatives are scattered and there is a strong need for their systematic documentation.

What the book is about

This book concentrates on smallholder agriculture in developing countries. This form of farming has been characterized as 'low external input agriculture' (LEIA), whereby the farming system relies almost exclusively on those resources available within the boundaries of the farm. Cover crops are, or have the potential to be, an important component in these complex, diverse, risk-prone and resource-poor farming situations. They may provide an appropriate technology to a wide range of circumstances, contributing to the goals of farmers, development institutions, natural resource conservationists and policy makers.

The book includes selected case studies from different countries within Latin America. These address key issues regarding cover crop integration in LEIA systems. A wide range of agroecosystems are covered by the case studies, proving that the information could be adapted for use in other regions. The case studies are based on the experiences and lessons learned from a variety of projects working in rural development and sustainable resource management. The key issues covered by the case studies are the following:

Cover crops in annual cropping systems (Honduras)

This study analyses the importance of indigenous cover crops in annual cropping systems from different agroecological zones within Honduras. The authors question past efforts to promote a single legume species that provides an extended mulch cover, instead of encouraging farmers to work with their traditional legume species, which contribute not only to soil improvement but also to family food security. The case is presented by CIDICCO, an international centre for information diffusion on cover crops.

Cover crops in perennial crops (Bolivia)

The use of cover crops in commercial plantations is commonplace. This case looks at ways in which this experience has been adapted to perennial crops within smallholder farming systems.

The role of cover crops in animal husbandry (Mexico)

This case study is based on the experiences of a collaborative project between a Mexican conservation NGO, the University of Yucatan and several rural communities living within the buffer zone of an important biosphere reserve in south-east Mexico. Cover crops have been promoted to reduce slash-and-burn activities. In order to increase the socio-economic viability of these cover crops they are being tried as an alternative source of animal fodder. Experimentation by the local communities is encouraged and facilitated for the adaptation of the innovations to the needs of individual families.

Natural resource conservation through the use of cover crop systems – soil improvement and conservation (Honduras)

The case study demonstrates the importance of soil conservation through cover crops in a mountainous area of northern Honduras. It shows how agricultural production and natural resource conservation are conflicting goals held by different stakeholders. The problems of deforestation and soil erosion leading to water contamination in the valley bottoms, and to poor crop performance, are being tackled through the introduction and promotion of cover crop systems. The experiences are outputs of a bilateral development project between Honduras and Germany.

Alternatives to slash-and-burn (Mexico)

The role of cover crops within a traditional slash-and-burn system in Yucatan, and their potential to contribute to a shift from rotational to permanent cropping systems are presented. Aspects of land tenure, communal resource use and governmental policies are highlighted to show the importance of addressing the wider socio-economic context within which an innovation is hoped to operate.

Diffusion aspects of cover crop based systems – *campesino*-to-*campesino* movement (Nicaragua)

The farmer-to-farmer diffusion approach is exemplified by an experience in the Bosawas region of northern Nicaragua. Methodological aspects are presented of how farmers' workshops and exchange visits contributed to the diffusion process of cover crop systems in villages bordering a forest reserve. The case is presented by the National Union of Agricultural Producers (UNAG), which has been working with this approach since the 1980s.

Applied research activities for agricultural systems improvement (Bolivia)

Approaches and methods for applied and participatory adaptive research into cover crops are described for a Bolivian research institution (CIAT). The elements of a wider strategy to address the problems caused by aggressive slash-and-burn agriculture are outlined, together with the role of cover crops in this situation.

Action–research with *campesino* farmers in south-east Mexico

Experiences with participatory research on cover crops in a traditional slash-and-burn system carried out by members of the University of Yucatan and several *campesino* communities are presented.

How to approach this book

General structure

Each chapter of the book addresses a different theme, and consists of a brief introduction and overview of the topic based on secondary information. The main body of the chapters are the case studies presented at the Merida Workshop, including the information obtained during the working group sessions. As a conclusion to each chapter, the main statements and discussion points are summarized.

However, the different aspects of cover crops cannot be treated separately from each other. The cross-cutting nature of the different aspects of cover crops becomes obvious from the case studies themselves. Each case study presented at the workshop addressed a key issue, but also provided valuable information on one or more of the other issues. We therefore decided to use the information from the case studies wherever it supports the development of the different themes. This explains why different parts of the same case study can be found in different chapters. The outline of the book is as follows.

Chapter 1 – The introduction gives some common definitions of cover crops, and then seeks to widen the perspective on the use of cover crops by presenting a range of potential functions and purposes of cover crops. It also considers the diverse regions in which cover crops might play an important role within the agricultural system and presents a table of the regions covered by the different case studies.

Chapter 2 – The use of cover crops in annual and perennial cropping systems is considered. Examples of traditional agricultural systems using cover crops are explored and a comparison is made with other cases where cover crops have been introduced. The different roles cover crops have in diverse systems are presented.

Chapter 3 – The value of cover crops as food, feed and forage is explored. Cover crops not only play a role within cropping systems, but are also used for human food and livestock feed. The main emphasis in this chapter is on feed and forage production because, as the Merida Workshop has shown, little attention has been paid so far to the issue of cover crop/livestock interaction and there is a clear demand from extension workers and farmers for more information.

Chapter 4 – Land husbandry with cover crops: this chapter considers the functions that cover crops have in terms of soil and water conservation. It also includes experiences where cover crops have been introduced to reduce deforestation caused by slash-and-burn agriculture.

Chapter 5 – Farmer experimentation and diffusion of cover crop innovations: based on two cases from Central America, this chapter presents the role of farmers in experimentation and diffusion processes of cover crop innovations. The experiences shared during the Merida Workshop showed that the introduction of new cover crops was always followed by further adaptation on farmers' fields and that horizontal communication across farmers and communities contributed significantly to the diffusion of such innovations.

Chapter 6 – Research strategies in Bolivia and Mexico for cover crop innovations: this chapter presents two case studies on how national agricultural research and extension (NARES) institutions carry out research in cover crop innovation development in collaboration with farmers. The aim of the chapter is to stimulate reflection on how NARES can improve their research for small-scale farmers in terms of increased uptake of innovation.

Chapter 7 – General discussion, conclusion and prospects. This chapter synthesizes the key issues and statements tackled in the previous chapters and provides functional and holistic definitions of the concept of cover crops as components of LEIA systems. It draws conclusions, and gives an overview of future strategies regarding cover crop integration into low external input systems. The conclusions reflect, and amplify, those reached by the Merida Workshop. Information gaps and key research issues are identified.

List of acronyms

AGRUCO	Agroecologia, Universidad de San Simon, Cochabamba (Bolivia)
C:N	carbon:nitrogen ratio
CIAT	Centro de Investigación Agrícola Tropical (Bolivia)
CIDICCO	Centro Internacional de Informacion sobre Cultivos de Cobertura
CONSEFORH	DFID project, Honduras
COSECHA	Honduras-based consultancy company
DFID	Department for International Development (UK)
DISE	Agricultural Extension and Systems Research Department
DM	dry matter
FMVZ	Facultad de Medicina Veterinaria y Zootecnia (at UADY)
FUNBANCAFE	DSE project, Honduras
GTZ	German Technical Cooperation
LEIA	low external input agriculture
NARES	national agricultural research and extension system
NARS	national agricultural research system
NGO	non-governmental organization
NRI	Natural Resources Institute (UK)
PCaC	*Programa Campesino-a-Campesino*
PROCAMPO	Government agricultural subsidy programme (Mexico)
RAFI	Rural Advancement Foundation International
SALT	Sloping Agricultural Land Technologies
SOM	soil organic material
SPSS	Statistical Package for the Social Sciences
UADY	Universidad Autonóma de Yucatán
UNAG	National Union of Agricultural Producers (Nicaragua)
USDA	United States Department of Agriculture

1

Introduction

Cover crops and low external input agriculture

Agricultural technologies developed to enhance productivity and to stimulate export-oriented agriculture are dependent upon external inputs (Conway, 1997). They have led to the intensification of agricultural production systems under favourable economic and ecological conditions, and they have contributed to the marginalization of resource-poor farmers unable to participate in the Green Revolution (Conway, 1997).

An important difference between external input-dependent agriculture and low external input agriculture (LEIA) is the degree of diversification in terms of production strategies and local resource use. Small-scale agricultural systems in the tropics are highly product diverse and location specific. For many resource-poor *campesinos* (*campesinos* are Latin American peasant farmers), technologies that have been developed within national and international research institutions are not appropriate, whether for economic, socio-cultural, managerial or environmental reasons. Reijntes et al. (1992) confirm the lack of appropriateness of many technology recommendations for location-specific contexts. They point out that in the context of *campesino* systems, farmers' heterogeneous needs and priorities have to be taken into account for technology development if the *campesino* sector is to be addressed successfully by research recommendations.

Cover crop innovations have become increasingly important in LEIA in many of the poorer agricultural sectors of the world. In the developing world many NGOs and farmer organizations have started to promote and develop cover crop technologies for LEIA systems (Buckles et al., 1998).

LEIA systems may be characterized by a series of features, which are important to consider with regard to cover crop integration.

Within LEIA systems there are several different types of agricultural systems with different potentials and options for the use of cover crops. Examples include shifting cultivation at the forest/agriculture interface, annual cropping systems, perennial cropping systems, mixed crop–livestock systems and extensive grazing systems. In the following chapters we present cases of the use of cover crops within these different types of agricultural systems.

Box 1 Functional and structural features of LEIA systems

- species diversity
- exploit a range of micro-environments differing in soil, vegetation and micro-climate
- high management requirement responsive to changing conditions
- production relies on local resources and family labour, and utilizes low levels of purchased inputs
- tend to rely on local varieties of crops, trees and animals
- integration of system components
- designed for stability and risk reduction rather than high production of individual components

Source: modified from Altieri (1991)

Cover crops have been components of traditional agricultural systems, but their use has been eroded by the promotion of farming practices based on external inputs and industrialized agricultural models (e.g. mono-cultures). The relative failure of Green Revolution practices in small-scale agriculture has led to a revision of thinking about the importance of diversity and integration of components in agricultural systems and a recognition of the contribution that cover crops can make to increasing productivity, income and the sustainable use of natural resources.

There is much to be learned from early accounts of the use of cover crops that may have application under present-day circumstances. A good example is the '*Mucuna* story', a plant introduced to the USA at the turn of the 20th century. The purpose at that time was to provide fodder for cattle, mules and horses, as well as to provide a soil cover in the monoculture plantations of citrus, cotton and maize (Buckles, 1994). The *Farmers Bulletin* of the United States Department of Agriculture (USDA) reported in 1918: 'Velvet beans (*Mucuna* spp.) are one of the most important crops of recent introduction and a determining factor in developing the livestock industry in the South. For a fertilizing crop the velvet bean is of greater value than the cowpea, as it makes a much heavier growth and is less expensive' (Tracy and Coe, 1918). Detailed research has been carried out to understand fully the potential of this plant. Data from the beginning of the last century are available on the impact of *Mucuna* on soil fertility and crop yields, on agronomic management practices, and on feed values. It was estimated in 1917 that the crop covered more than 5 million acres in the USA. Less than 70 years later the same plant has regained importance, this time not in the USA but in different developing countries around the world. At present, *Mucuna* is considered to be the most popular cover crop or green manure species (Flores, 1994).

Towards a definition

Before we delve further into details of the use of cover crops we should first look at the definition of the concept. Different terms for the cover crops are used in the literature. *Live mulch, organic mulch, green manure, slash/mulch* systems and *cover crops* are those most commonly found. Several definitions have been established by various authors. For instance, Kiff et al. (1996) define cover crops in the following way:

Box 2 Definition of cover crops

'temporary or permanent, live, non-woody, vegetative soil cover grown within rain-fed annual or perennial cropping systems'
Source: Kiff et al. (1996)

This definition specifies the type of cover provided by cover crops as well as the systems where they can be found. But it lacks an explanation of the function of cover crops within an agricultural system. Another attempt to define cover crops is as follows:

Box 3 Definition of cover crop concept and functions

Cover crops are 'herbaceous crops grown to create a favourable soil microclimate, decrease evaporation and protect soil from erosion; also to produce biomass that can be used as forage and to improve the soil.'
Source: Bayer and Waters-Bayer (1998)

The latter definition stresses the different functions of cover crops without specifying where these cover crops can be found. Some sources emphasize that cover crops always refer to perennial plantation systems.

Flores provides a definition that distinguishes between *cover crops* and *live cover* – the former for perennial systems and the latter for annual cropping systems.

3

Box 4 Cover crops or live cover

Cover crops, if used correctly, refer to those crops established simultaneously in perennial plantations. They are usually not incorporated into the soil. Other creeping species are established within grain crops for weed control, erosion control and nitrogen fixation. These are generally called *live cover*.

Source: Flores (1993)

Somewhat contradicting the previous definition is the following provided by Thurston:

Box 5 Cover crops as green manure

'Cover crops refer to those crops planted for green manure'. The same author suggests later that 'cover crops are any crops grown to produce soil cover, regardless of whether they are later incorporated. They are used to cover and protect the soil surface, although they may be turned under as green manures.'

Source: Thurston (1994)

In the praxis of small-scale farmers around the world, cover crops are used in many different ways. Enquiring into and understanding farmers' strategies in using cover crops within their small-scale agricultural systems will provide insight into their definition of the concept. The lessons learned from the field show that the understanding of cover crops as merely ways to improve soil fertility is misleading, and that this perspective has somehow limited the exploration of their full potential.

Cover crops can be described, or defined, in relation to their biological classification, the functions that they perform, the purposes to which they contribute (for different stakeholders, including farmers, development agencies, donors, policy makers, natural resource conservationists), and the physical and socio-economic context in which they are used.

During the following chapters we present information and practical field experiences that explore the concepts behind the use of cover crops and identify the functions of cover crops and the purposes to which they contribute. This information helps to develop a more holistic definition of the concept of cover crops, which is given in Chapter 7. An agroecosystems perspective is taken to guide the analysis.

Functions and purposes of cover crops in agriculture

The distinction between function and purpose in this section is just an introduction to a discussion that will form an integral part of this book. While *function* is understood to be determined by the potential that is inherent in a certain technology or system component, *purpose* is defined as a person's decision on why and how this technology is used. This decision will be determined by individual preferences, needs, information access, etc. and will vary between different stakeholders, as the example in Box 6 shows:

Box 6 Why establish *Mucuna* in hillside agriculture?

A *campesino* in the tropical highlands of Chiapas decides to establish *Mucuna* as a cover crop in his fields, which are on steep mountain slopes. He needs extra fodder for his two cows and he has heard about *Mucuna* as a high-protein forage. For him the purpose of sowing *Mucuna* is to produce more fodder for his animals. An extension worker in the region provides the *campesino* with *Mucuna* seeds, as his purpose is to promote measures for soil erosion control in order to increase crop yields and to reduce the siltation load of a nearby river.

This example highlights a series of important points. First of all, it shows that by establishing *Mucuna* in the field these different purposes might be achieved simultaneously, and at the same time *Mucuna* can fulfil some additional functions such as soil fertility improvement. It also shows that the contribution of *Mucuna* occurs not only at field level (e.g. soil improvement, yield increase), but also at farm level (animal fodder) and even at watershed level (decrease of siltation). The range of functions of different cover crops allow stakeholders to select those that are appropriate to their circumstances and to prioritize them according to their objectives and purposes.

Cover crops within an 'envelope' of conditions

Cover crops can be found across a wide range of different conditions. The selection of an appropriate species and the way it contributes to the agricultural system depend on it being able to satisfy most of the requirements contained in an 'envelope' of conditions. These physical and socio-economic conditions might include:

- temperature, rainfall and daylength

- specific physical conditions (e.g. waterlogging, wind, etc.)
- soil conditions
- competition and complementary effects with main crop
- ability to perform required functions
- complementary to other farm activities
- level of intensification of the agricultural system
- land and cash resource availability.

Most research attention so far has been paid to biophysical conditions. Temperature, rainfall pattern and altitude strongly influence the phenotypic development in annual crops. To meet the requirements for cover crops grown under a variety of climatic conditions, a wide range of species has to be considered. Keatinge tested a range of different cover crop species and varieties in terms of their performance (Keatinge et al., 1996). More detailed information on climatic requirements can be found in the NRI annotated database on cover crops (Kiff et al., 1996).

Table 1 lists some species suggested for high and cold temperature conditions. The list exemplifies the diversity of cover crops used.

Table 1 Examples of cover crop species for different temperatures

Species for high temperature conditions	Species for cold temperature conditions
Canavalia spp.	Vicia faba
Mucuna spp.	Lupinus mutabilis
Dolichos spp.	Vicia spp.
Stylosanthes spp.	Trifolium spp.
Vigna spp.	Pisum sativum
Phaseolus spp.	Phaseolus spp.
Desmodium spp.	Brassica spp.
Pueraria spp.	Sinapis alba
Centrosema spp.	Raphanus sativus
Crotalaria spp.	Phacelia tanacetifolia
Cajanus cajan	
Tephrosia spp.	

Source: Keatinge et al. (1997)

The list shows the wide selection of different cover crop species adapted to different climatic conditions. This variability of species has allowed the use of cover crops across a wide range of geographical regions. Cover crops are established in agricultural systems from humid to arid, temperate to tropical and from lowland to highland environments. However, the main functions they perform in the different regions, and the management strategies applied by farmers, vary widely between regions.

Industrial agriculture, located in *temperate regions*, can be characterized by a high degree of intensification and high levels of external input dependency. This type of agriculture has created problems in terms of environmental contamination. Cover crops are increasingly being used in these systems to reduce the level of inorganic inputs like herbicides and fertilizer. They also play an important role in reducing nitrogen leaching during off-season periods. *Vicia* spp. and *Trifolium* spp. as well as non-leguminous species (*Brassica* spp.) are common in temperate regions (see Decker et al., 1994).

In temperate regions of the developing world, cover crops have very different functions from the ones described above. They contribute to the production of food, feed and forage and they protect the soil from erosion and nutrient losses. An example is the use of a traditional cover crop, *Medicago hispida* (Garrotilla), which is associated with potatoes or wheat in the highlands of Bolivia (see Box 2.10).

In *semi-arid regions* cover crops can play an important role in soil water conservation and wind erosion control. They are frequently established during the rainy season together with a main crop, e.g. maize or sorghum. Drought-tolerant species like *Canavalia ensiformis* provide soil cover for two or three months after the rains have ceased, whereas in other cases the soil would remain bare until the next cropping season. Another example is that of cover crops such as *Mucuna pruriens*, *Stylosanthes hamata* and *Voandzeia subterranea* (bambara groundnut), which provide ground cover, human food and animal feed (Kiff et al., 1996).

In the *tropical lowlands* cover crops play important roles in soil fertility management and the intensification of agricultural systems. As these regions are generally more densely populated than hillside areas, pressure on land is high. Due to the climate the decomposition of organic matter is fast and soil fertility can decline quickly. Farmers can seldom afford to leave part of their fields out of production. Therefore, cover crops are often required to fulfil several purposes, e.g. food/feed production and green manure production for the soil. *Phaseolus* spp. and *Vigna* spp. are examples that can combine both of these purposes.

Many regions of the *tropical highlands* are characterized by their remoteness from markets and economic development (e.g. the Andean region). Cover crops, once introduced, are a local resource reducing dependency on external inputs. They are therefore particularly appropriate to remote communities where communication and access to inputs is limited, and also to those households with limited financial resources. Cover crops are important for their use in soil conservation and fertility management under those conditions.

The case studies presented in this book deal with different geographical regions and give examples of the different functions and purposes that cover crops perform under diverse conditions. Table 2 provides an overview of the case studies presented in this book, based on their geographical location.

Table 2 The geographic locations covered by the case studies

Geographic location	Case study country and key characteristics of the region	Main functions and purposes of cover crops
Temperate regions	**Bolivia (Cabecera de Valle):** altitude between 2800 and 3200 m, humid microclimates, traditional agricultural systems	system diversification, feed source, soil improvement
Semi-arid areas	**Honduras (Choluteca), and Mexico (Calakmul region):** dry season, water stress, marginal soils	soil water conservation, food/feed production
Humid tropical lowlands	**Nicaragua (Bosawas reserve), and Bolivia (Santa Cruz department):** fast organic matter decomposition, slash-and-burn agriculture, rapid weed growth	soil fertility management, land-use intensification, weed management
Tropical highlands	**Honduras (Merendon region):** extreme temperature differences, shallow soils, remoteness	erosion control, sustainable resource use

2

Cover crops in annual and perennial cropping systems

Introduction

This chapter sets out how different types of cover crop are integrated into annual and perennial cropping systems. Examples from Honduras and the highlands of Bolivia show how traditional cover crops can contribute to crop and crop/livestock agricultural systems, while a case study from Mexico demonstrates that the introduction of a cover crop can be key in making the change to a more sustainable type of agriculture. Experiences from lowland Bolivia are used to highlight the potential of cover crops in smallholder perennial cropping systems.

Crop/cover crop associations

The nature – growth habit, shade tolerance, competitiveness – of the cover crop to be associated with the main crop, as well as the purpose for using the cover crop, influence the selection of possible crop combinations. It is important that cover crops are complementary to, rather than competitive with, the crops they are associated with, both in time and space. It is assumed that farmers would give priority to the main crop in moments of conflicting requirements, although medium- to longer-term effects might cause a farmer to compromise short-term benefits. An example of this is that many farmers tolerate the difficulties in harvesting maize in a maize/*Mucuna* association (as long as the yield of maize is maintained) because of the longer-term beneficial effects on weed and fertility management (Gündel, 1998).

The development of a successful crop/cover crop association depends on the selection of compatible species and varieties, and the use of appropriate management practices, i.e. planting dates, spacing and planting patterns, cutting to reduce competition, methods for elimination of the cover crop when no longer required, mulch management and the mitigation of negative effects such as fire risks, and above- and below-ground competition.

Competition between crop associations can be for:

- light, e.g. some cover crops grown with pineapple inhibit the flowering of pineapple due to their shading effect
- nutrients, e.g. it is thought that the root systems of *Arachis pintoi* and peach palm (*Bactris gasipaes*) occupy similar soil space, leading to poor growth of peach palm (CIAT/NRI, 1997)
- water, that may limit cover crop options in those areas or seasons when soil moisture limits main-crop growth.

The choice of cover crop will depend on the size, vigour, growth pattern and stage of the main crop, the climatic and edaphic conditions, labour availability, and knowledge of, and access to, alternative species.

The ideal characteristics of a cover crop (after Binder, 1997) might include:

- adaptation to poor or degraded soils
- low seed cost
- seed (or vegetative planting material) easily managed and multiplied
- multiple utilization (food, feed, medicine, etc.)
- reliable germination without the need for seed treatment
- rapid emergence and soil cover (good competition with aggressive weed species, without becoming a weed itself)
- vigorous root system able to penetrate compacted soils, improve infiltration and recycle nutrients from depth
- root system does not compete with the main crop for water or nutrients; aerial component does not compete with the main crop for light, or interfere with its growth, management or harvesting
- growth habit and cycle adapted to that of associated main-crop(s)
- high nitrogen-fixing capacity over a range of climatic and soil conditions
- degradation rate and carbon:nitrogen (C:N) ratio of residues compatible with the needs for soil cover, soil organic matter enhancement and nutrient demand of associated main crop(s)
- low susceptibility to pests and diseases
- non-hosts to pests and diseases of the main crop
- tolerance to key conditions (drought, heat, cold, waterlogging, shading, etc.).

Cover crops can be identified (Binder, 1997; Kiff et al., 1996) for tropical (e.g. *Mucuna pruriens, Canavalia ensiformis*) and temperate (e.g. *Vicia villosa, Lupinus mutabilis*) conditions, and some have specific tolerance of waterlogging (e.g. *Sesbania sesban, Aeschynomene* spp.) drought (e.g.

Cajanus cajan, Canavalia ensiformis, Macroptilium atropurpureum) or shade (e.g. *Calopogonium mucunoides, Centrosema pubescens* and *Desmodium ovalifolium*).

Many are tolerant of poor soil fertility, and some are able to tolerate high-aluminium levels in acid soils (e.g. *Vigna unguiculata, Lablab purpureus, Arachis pintoi, Stylosanthes guianensis, Centrosema* spp.).

Care has to be exercised when planting out of season, or when taking seed from one latitude to another, as some species are sensitive to photo-period, and may not set seed unless daylength is appropriate. Short-day genera include many of the tropical leguminous cover crops (e.g. *Neonotonia, Canavalia, Mucuna, Vigna, Lablab, Centrosema, Desmodium*).

Types of cover crop

Diverse types of cover crops are used in annual and perennial cropping systems throughout the tropics and sub-tropics. The majority are legumes, but some gramineous species are used (e.g. *Avena negra, Paspalum conjugatum*).

Cover crops can be grouped into three main types according to their growth habit which strongly influences the management strategies and suitability of different cover crops in farming systems.

Trailing annual species like *Mucuna pruriens* and *Vigna* have the advantage of establishing cover rapidly. Figure 1 demonstrates the rate at which different species cover the soil surface. *Mucuna pruriens* is an annual, trailing legume; *Canavalia ensiformis* is an erect annual species; *Cajanus cajan* is an erect biennial, and *Arachis pintoi* is a prostrate, perennial species.

The vigorous growth of trailing legumes such as *Mucuna* and *Puraria phaseoloides* requires careful management in order to reduce unwanted competition with the associated food or cash crop.

Arachis pintoi (also known as perennial- or forage-peanut) is an example of a prostrate, perennial cover crop. Establishment of this class of cover crop tends to be slow (see Figure 1) but, once established under the right conditions, they can provide a long-term, low-maintenance cover that controls weeds and reduces soil erosion. On-farm validation trials combining *A. pintoi* with a range of perennial crops, carried out by CIAT, Bolivia, revealed that farmers did not like its slow establishment and they also complained about poor germination (often caused by poor seed quality or wet planting conditions). Once it has established a dense ground cover, it is reported to have a positive impact on weed control and soil conservation. Management requirements during the cropping cycle are less than for vigorous climbing covers because of its non-aggressive growth habit.

Shrubby legumes include *Cajanus cajan* (pigeon pea), which offers a wide range of different varieties that can be annual, biennial or perennial.

Figure 1 Development of different legume species in perennial agroecosystems

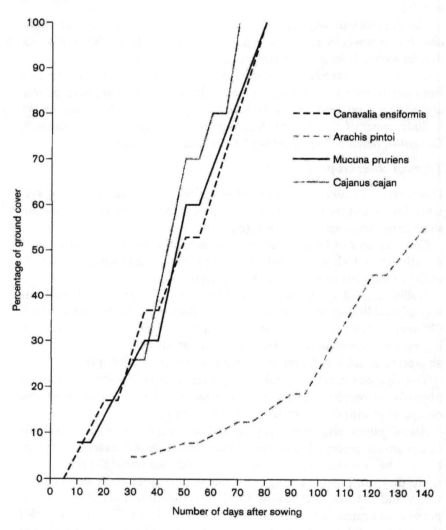

Cajanus cajan grows as a shrub with a strong taproot. It needs sufficient water during the first two to three months of establishment, but subsequently can be considered as quite tolerant of drought stress. The leaves form an excellent mulch for the soil. The erect growth habit makes it easy to manage, and the peas are a good source of protein for human diets. Positive impacts of maize/*Cajanus cajan* intercropping are reported from Africa as well as from the Caribbean.

Another example within this category is *Canavalia ensiformis* (Jack bean). There are some trailing forms, but others, especially those coming from Venezuela, present a non-climbing, erect growth. *Canavalia ensiformis* is cultivated on a small scale throughout the tropics as a cover crop, green manure, forage, and as a grain legume for animal and sometimes for human nutrition (Kessler, 1990). It is a fast-growing plant (see Figure 1), which establishes easily and quickly from large seeds, providing a good, early soil cover. *Canavalia ensiformis* is a versatile, easily managed cover crop which can be successfully established from sea level up to 1800 m, and in climates from warm/temperate to hot/moist and hot/arid (Keatinge et al., 1997). Species such as *Canavalia ensiformis* have a large number of landraces with different characteristics (Kessler, 1990). To date, little selection has been made to match landrace to niche need.

Very brief descriptions of the cover crops mentioned in this book can be found in Appendix 3. Publications providing further details of cover crops by species include Kiff et al., 1996 (in English); Monegat, 1991 and Binder, 1997 (in Spanish), and Caligari, 1995 (in Portuguese).

Cover crops in annual cropping systems

Most experiences of cover crops in Central and Latin America are reported for annual staple-crop (e.g. maize) production systems. However, there exists a range of possible crop combinations within annual cropping systems. Table 3 summarizes the experiences presented during the Merida Workshop. Some of these crop combinations are traditional practice, whereas others are the result of recent promotion activities.

Table 3 Crop associations between main crops and cover crops

Main crop	Cover crop	Country
Maize	*Canavalia ensiformis*	Mexico
	Vigna spp.	Honduras
	Dolichos lab-lab	El Salvador
	Mucuna pruriens	Guatemala
	Phaseolus coccineus	Nicaragua
Maize and millet	*Canavalia ensiformis*	Honduras
	Mucuna pruriens	Nicaragua
	Phaseolus vulgaris	El Salvador
	Vigna unguiculata	

Millet	*Phaseolus vulgaris*	El Salvador
		Honduras
Squash and maize	*Vigna* spp.	Mexico
	Phaseolus vulgaris	Honduras
Maize and tomato	*Vigna* spp.	Mexico
	Canavalia ensiformis	
Chillies	*Canavalia ensiformis*	Mexico
Rice	*Mucuna* spp.	Belize
	Dolichos lab-lab	
	Canavalia ensiformis	
Potatoes, oats, barley	*Medicago hispida*	Bolivia (sub-tropical/
		temperate)
Vegetables	*Cajanus cajan*	Honduras

Source: Merida Workshop, 1997

Traditional cover crop uses

The use of cover crops within agricultural systems is not a 'new' practice: the concept can be found in many traditional, small-scale agricultural systems. During recent years renewed attention has been paid to these systems by farmers, researchers and development agents searching for sustainable agricultural technologies. Cover crops are now considered to have potential for improving agriculture in marginal regions, and are promoted as an 'innovation' for sustainable land use. In this section we present some examples of cover crops as components of traditional agriculture to show that there already exists important knowledge and experience in rural communities around the world, which constitutes an important source of information for the future development of alternative management strategies.

The term 'traditional' is used here to describe practices that are appropriated by local farmers. They are not necessarily ancient traditions but rather local practices, which have been developed as part of farming activities. Farmers' knowledge is dynamic, and spatially and socially non-uniform. The following anecdote, which is borrowed from a West African context, presents a good example of how external innovations become part of local knowledge.

Box 7 An anecdote from West African rice farmers

A conversation between a researcher and a rice farmer in West Africa:
Researcher: 'Why don't you use fertilizer?' Farmer: 'You mean puumoi [European] fertilizer, in bags?' Researcher: 'Why, is there another kind?' Farmer: 'Yes, of course...Mende fertilizer.' Researcher: 'What kind of fertilizer is that?' Farmer: 'You know the plant magOni? Look, here is some; it grows as a weed in farms but after it has grown the soil has strength again. That is Mende fertilizer.'

Later on the researcher checked magOni (*Calapogonium mucunoides*) against the herbarium specimen at Njala. There, the note on the file reads 'introduced to Sierra Leone in 1938 as a nitrogen-fixing plant for green manure experiments'. The farmer's 'Mende fertilizer' could have been in Mogbuama no more than 45 years. On the next visit the researcher told the farmer this. His answer was: 'I didn't say Mende fertilizer was old, or that we invented it...but we know about it and you don't; that's what makes it Mende.'

Source: Richards, 1995

Cover crops have the potential to improve soil fertility and structure, and control invasive weeds, and can therefore contribute to the replacement of fallow systems (including slash-and-burn systems) by associations between cover crops and annual or perennial crops. Eliminating burning reduces the losses of organic matter and nutrients associated with this practice. In no-burn fallow systems, the use of cover crops can play a part in the intensification of land use.

Case studies of local cover crop uses in annual cropping systems

Two experiences from Honduras (prepared and presented by CIDICCO, 1997) and a third from the Bolivian highlands (AGRUCO, 1997), are presented below. These describe local cover crop uses in different annual cropping systems.

Box 8 Lowland cover crop systems in Honduras

In the southern region of Honduras *campesinos* traditionally grow basic grains and *Vigna* in a mixed cropping system. Despite the advice from extension services to establish maize monoculture cropping systems, this local cover crop system has survived until today.

Some facts about the region: The Choluteca region is one of the most densely populated areas of Honduras (86 inhabitants per km²). Rainfall varies

---

between 1400 and 1600 mm per year, but the distribution is very bad. In some years most of the rain occurs within three weeks. Soils contain plenty of stones and are low in organic matter. The natural vegetation is a dry forest, which has been cut down in most areas due to population pressure. Yields are low in the region, resulting in poor food security. Main crops are millet and sesame, followed by maize. Livestock production is an additional activity.

History of the mixed cropping system: Campesinos report the origin of the system to date back to the first settlers coming to the region. They believe that the *Vignas* were of local origin and have been selected for intercropping with basic grains. Different types of *Vigna* spp. (mostly *Vigna unguiculata*) were used within this system, and are known as *Alacines*, *Cuarentanos*, *Media vara* and *Pochote*. These can still be found in the region, despite the fact that there exists no national market for *Vignas* in Honduras. They are grown for family food, as well as contributing to soil fertility management and providing forage for livestock, which are commonly brought into the fields after harvest.

Management aspects of the system: With the start of the rains in April the *campesinos* sow maize, millet and *Vigna* with a planting stick into the soil. Maize, millet and *Vigna* are sown in the same planting hole. Often different types of *Vigna* are planted within the same field. As they have different growth cycles, the harvest of beans can begin after 40 days with a short cycle variety, while other *Vigna* need four months to mature. During the first month manual weed control is done twice. The next activity is the harvest of the beans, which are eaten green, ripe or dry. The yields vary between 10 and 40 *quintals* per *manzana* (1 *quintal* is equivalent to 100 pounds, and 1 *manzana* equals 7000 m²), depending upon the management practice. Low yields are obtained under traditional broadcasting, whereas better yields are obtained in improved tillage systems. After the harvest is finished, animals are brought into the field for grazing or the hay is cut, dried and then carried to the homestead, where it is fed to the animals.

Diffusion aspects of the technology: Although the system has been shown to be beneficial for the region, the diffusion to other regions has been limited. This is due to the fact that national research and extension services have neglected the importance of these traditional systems, and have instead promoted the use of high-yielding bean varieties, which required the adoption of technology packages including fertilizer and herbicides. These technologies have not been suitable for small-scale *campesino* systems as they are too expensive and risky to adopt. Unfortunately they have contributed indirectly to a process of devaluation of local species and technologies, as these have been regarded as being inferior to the improved technologies.

Impacts of the Vigna system:

- improved level of food security due to the reliability of the *Vigna* spp., which present a high drought resistance compared to other crops of the region
- the early-maturing types provide human food during a season in which food is scarce
- improved forage production for the feeding of livestock
- weed control through *Vigna* cover
- soil fertility and soil organic matter improvement
- compatibility between *Vigna* and other local staple crops

Source: case study prepared by Raúl Alemán, Myriam Paredes and Norman Sagastume, CIDICCO, Tegucigalpa for the Merida Workshop.

Issues highlighted by the *Vigna* case study

This case study from Honduras highlights some crucial issues with regard to the use of cover crops within local agricultural systems. The envelope of conditions in which the cover crops are grown appears to be not particularly favourable: high population density, limited land resources, seasonal water shortages, poor soils and no market potential for the cover crops grown. Additionally, external extension activities have tried to promote maize monocropping systems and high-yield bean packages, obviously not taking into account local sources of information and the objectives and needs of local farmers. Despite these factors, the local technology has survived.

The success of the local practice can be summarized as follows:

- *Vigna* can be used for multiple purposes within the farming system.
- The use of local *Vigna* varieties is appropriate to farmers' socio-economic conditions.
- *Vigna* varieties are adapted to local conditions.
- The cover crops are compatible and complementary with other components of the agricultural system (crops and livestock).

Regional and national research and extension services have neglected the importance of this traditional system, which is manifested in the limited diffusion between communities of the region.

A further case study, also provided by CIDICCO, demonstrates the locally-generated use of cover crops under a different set of conditions. The case study again comes from Honduras, but this time from the hillside areas, where the farming system has to cope with a different set of problems.

Box 9 The Chinapopo system of the highlands of Honduras

Campesinos from Honduras, and elsewhere in Central America, are populating marginal hillside areas, as modernized agriculture and insecure land rights have displaced them from higher-potential areas. The fragile environment of the tropical highlands requires production systems to adapt to their environmental and socio-economic limitations.

History of the system: For very many years the cultivation of Chinapopo beans (*Phaseolus coccineus*), in combination with other staple crops, has been an important agricultural practice contributing to family food security. The Chinapopo beans originally came from the highlands of Mexico. The beans are adapted to highland conditions, allowing them to be grown in areas between 1400 and 2800 m.

Some facts about the region: The highlands of Honduras are between 1400 and 2000 m altitude. Temperatures are between 15 and 20°C and the rainfall averages 1500–2000 mm/year. The soils are deep, composed of volcanic material and contain a moderate level of organic matter. The slopes range between 15 and 70 per cent and the soil is susceptible to erosion. Apart from subsistence farming, coffee plantations can be found in the region – where *campesinos* work temporarily as labourers.

Subsistence production system: The *campesinos* of the highlands concentrate their activities around the sowing of maize and beans (*Phaseolus vulgaris, P. coccineus*), which are the staple crops for their families. The production cycle of this system is nine months and the most important factor influencing the success of the system is the rainfall. As the Chinapopo bean has a climbing growth habit, maize is important to support the bean plant. Chinapopo bean can be considered a perennial cover crop. The plant produces tubers, which can remain in the soil for several years and which produce new shoots every year. However, the *campesinos* sow Chinapopo every year, as this results in bean harvests at different times. Plants that grow from the previous year's tubers mature several weeks ahead of the ones that are sown annually with the maize. Chinapopo is consumed green or dried. After the harvest the remaining plant material is grazed by animals, or carried to the homestead for use as livestock feed for stalled animals.

Benefits of the Chinapopo system:

- the system is socio-culturally and ecologically adapted in the region, as it has been developed by the people

- it allows simultaneous production of maize and beans and decreases the risk of total harvest loss
- Chinapopo is one of the few sources of protein in the region
- Chinapopo increases soil fertility and organic matter content of the soils (preliminary observation shows 100 rhizobial nodules per plant root of newly established plants)
- reduced soil erosion and controls weeds
- improved forage availability and quality.

Some limitations:

- no economic potential/no market for Chinapopo
- lack of knowledge among *campesinos* of the additional benefits of Chinapopo for maintenance of soil productivity
- recently introduced vegetable growing is affected by the tubers of Chinapopo
- Chinapopo can reduce maize yield if not managed appropriately
- development organizations of the region have not recognized the benefits of the system and consequently do not promote them.

Source: a case study prepared by Raúl Alemán, Myriam Paredes and Norman Sagastume, CIDICCO, Tegucigalpa, for the Merida Workshop.

Issues highlighted by the Chinapopo case study

Although the environmental conditions and species used are different from those described for the lowland farming case study, the purposes required by farmers of the cover crops are similar in both cases. One important purpose is to increase food security by diversifying the range of crops grown and their potential harvest period. The other main purpose is to improve forage quality and quantity for livestock kept in the system.

The fact that farmers take the animals to their fields to graze the crop residues after harvest, or carry the residues to their homesteads where the animals are kept, suggests that either they are not aware of the potential contribution of these plant residues to soil fertility and erosion control, or they value their contribution as livestock forage higher than the potential long-term contribution to maintaining soil productivity.

Despite long-term local experiences of farmers with different cover crops, knowledge gaps exist regarding the potential functions of these technologies – on the part of farmers, as well as on the researchers' side.

The third case study, reported from the Bolivian highlands, points out the divergence of knowledge between local farmers and researchers.

Box 10 More highland experiences – The Garrotilla (*Medicago hispida*) in Bolivia

The context: *Campesinos* living in the Cochabamba valley have developed a traditional cover crop system which integrates wheat, oats, maize, barley, potatoes and some other crops with a leguminous plant, locally known as Garrotilla (*Medicago hispida*). Garotilla can be found between 2800 and 3400 m. This zone receives 800–900 mm of rainfall per year, and the average temperature is about 18°C. Garrotilla is a locally-occurring legume that has been integrated by the *campesinos* into the local farming system.

Management practices: Garrotilla is associated in the traditional cropping system with either basic grains, tuber crops or other edible legumes. Its vigorous establishment contributes to the control of weeds within the field. The vegetation cycle of Garrotilla is between six and nine months. It is usually established with the onset of the rains in October/November and produces seeds between February and May. Although Garrotilla does not provide human food, it contributes to the livestock component of the farming system, as the main use of the plant is as forage. It is either cut and carried, grazed once the harvest of the main crop has been completed, or dried and stored as hay. Garrotilla is also cultivated in fallow areas where it is either grazed by livestock or cut and carried.

Benefits of the cover crop technology
The plant provides several benefits to the *campesino* production system:

- no special management requirements; no seed costs as the plant is distributed in droppings by animals fed on Garrotilla
- produces a high-quality forage with 17.3 per cent protein and with good storage qualities, which provides fodder in critical seasons
- erosion control (wind and water)
- provides shade for the soil, which contributes to water conservation and organic matter accumulation
- capacity to fix nitrogen and to pump phosphorus from deeper soil levels.

Disadvantages:

- limited to elevations below 3400 m
- high requirements for soil nutrients
- reduces tuber crop development
- high demand for water, which precludes establishment in sandy soils.

Source: Poster presented by Elvira Serrano, AGRUCO, Universidad San Simon, Bolivia at the Merida Workshop (Anderson et al., 1997).

Issues arising from the Garrotilla case study

This case from Bolivia again shows the multiple purposes and functions that a cover crop may assume in a certain system. Knowledge of functions like nitrogen fixation and phosphorus pumping are not accessible to farmers through their day-to-day experience and observation. These are aspects that require more formal research to understand them thoroughly.

There are many examples of local cover crop uses. For example, the use of lupins (*Lupinus mutabilis*) was a common practice in pre-Inca agriculture in the Andes. The plant can be established up to an altitude of 3000–4000 m, which makes it an important component of high-altitude farming systems. Recent activities in Bolivia have promoted this traditional plant in various Andean communities. The objective is to increase staple food production by integrating lupins as a cover crop prior to potatoes (Thurston, 1997).

The case studies presented fulfil the purpose of highlighting the complexity of cover crop integration in terms of potential functions and farmers' purposes. Key aspects of cover crops in local systems are further discussed below.

Observations from the case studies on traditional cover crop/ annual crop associations

Diversity and complexity of the farming system

The farming systems presented above can be considered as complex in terms of crop–crop and crop–livestock interactions. They are based not only on one crop in combination with a cover crop, but on a range of crops, all of which have specific functions in terms of time and space within the overall system. Different crop species and varieties are combined in such a way that resource use is optimized, and the harvest period is stretched over a longer period by using short, medium and long-cycle crops.

Crop–livestock interaction

Low external input farming systems tend to be multi-component, containing a range of crops and livestock. In these systems cover crops are an integral part of the system, contributing to both crop and livestock components.

Multipurpose function of cover crops

The cover crops *Vigna*, Chinapopo (*Phaseolus coccineus*) and Garrotilla (*Medicago hispida*) perform multiple purposes and functions within the systems described. One main purpose of the *campesinos*, which is shared among the different cases, is the provision of food and/or feed. *Vigna* and Chinapopo form an important part of the local diet and can be consumed during various stages of their growth cycle (green, ripe, dry). As a source of protein they are important to the *campesino* families and also to the livestock kept within these systems, as they consume the by-products after harvest. Garrotilla is an exception, as its main purpose is as a livestock feed; it is not consumed by humans. Apart from this, the three species have additional functions within the systems, which might or might not be recognized by the farming families (weed control, water conservation, soil conservation and soil fertility improvement).

Location specificity of farmers' knowledge
The fact that these systems have been developed over long periods under local circumstances has resulted in very specific practices based on the individual knowledge of farmers and the availability of crop species in different micro-regions. Many species used in traditional cover crop systems are reported by the farmers to have their origin in the region, and have been adapted according to specific needs. Others have been introduced in the past and have been appropriated by the farmers, who have again developed them according to their criteria.

Temporal and spatial arrangements
Cover crop use varies temporally as well as spatially. As an agricultural system develops through different stages of intensification, so the function of cover crops may change. Farming systems are dynamic, which leads to new cover crop niches emerging within them. This process can be assisted if farmers are aware of the alternative components (including cover crops) that might be introduced to suit the new situation.

Promotion of 'new' cover crop technologies

Cover crops can contribute to goals that are important to policy makers, natural resource conservationists, development institutions and donors, as they are considered to be a low-cost, low-input and sustainable technology, which contributes not only to agricultural production but also to environmental conservation. At present, the popularity of cover crop technology is high among external stakeholders, which has led to many activities around the world aimed at the promotion of cover crops.

In this section the focus is on cover crop technologies that are promoted by different stakeholders in order to provide alternative management strategies seeking to cope with the diverse problems faced by farming families and the environment.

A study of existing literature and project activities clearly shows that a limited number of cover crop species are being tested and disseminated. This is as true for research activities as for promotion at village level.

The popularity of *Mucuna*

Mucuna pruriens is the most popular cover crop (particularly in the humid tropics) at present, but there are concerns about the sustainability of *Mucuna* use. Seeds are passed from one project to another, leading to a narrow genetic base. Until now, evidence of pests and diseases have seldom been reported, but other experiences (such as that of the *Leucaena* psyllid) have taught us some lessons on the danger of over-reliance on a small number of species. This disquiet is shared by many people. 'In fact there is some concern that the great attention that the scientific community has focused on this species [*Mucuna*] may have introduced an element of bias. Even more importantly, nearly all studies have centred on fertility, omitting the fact that the use of this plant is really part of a production system which combines social, economic, agricultural and ecological considerations' (Flores, 1994).

However, there must be good reasons why *Mucuna* is among the most popular cover crop species, not only with development workers and researchers but also with farmers. The following list gives some reasons for the widespread diffusion of this species:

- ease of dissemination; *Mucuna* seeds well and the large seeds keep their viability for several years
- rapid cover (see Figure 1), leading to good weed and erosion control
- high amount of organic matter
- good growth under a wide range of soil conditions
- good nodulation, even under acid soil conditions, leading to excellent contribution to soil fertility
- few pests and diseases to date
- easy to eliminate, reducing competition with subsequent crops
- potential for livestock feed (see Chapter 3)
- well known to development institutions.

Case studies: Introduced cover crops in annual cropping systems

The first example of an introduced cover crop technology comes from Yucatan, Mexico, where several NGOs promote a technology known as 'minimum tillage' or *labranza minima*. In industrialized countries minimum tillage is promoted as an alternative to high levels of mechanization that have resulted in soil degradation. For mechanized agriculture in developing countries it could also serve the same purpose. However, many smallholder *campesino* systems in Latin America have traditionally been zero-tillage systems (e.g. slash-and-burn systems).

In slash-and-burn systems that rely on an extended bush-fallow period (including *milpa* systems of Central America, *barbecho* systems of Bolivia and *capuera* systems of Brazil) the main functions of the bush-fallow are to recuperate soil fertility and structure, and reduce the incidence of aggressive weeds. Generally this requires a minimum fallow period of some eight years. Under some circumstances cover crops may be able to fulfil the functions of a bush-fallow in less time, and without the loss of nutrients associated with burning. The aim of the NGOs in Yucatan in promoting minimum tillage is to provide an alternative to the local slash-and-burn agriculture, which is associated with problems of declining soil fertility, weed invasion, low maize yields, and deforestation. This technology was promoted in Central America during the 1980s and has been widely adopted by farmers. The recommendations regarding implementation and management practices have remained very similar over time, although farmers have made various modifications to the original technology.

Box 11 'Minimum tillage' in Yucatan

Some facts about the region: The Yucatan peninsula is located in the southeast of Mexico. The altitude of the peninsula ranges between 0 and 300 m and the average temperature is around 24°C. The agricultural cycle is divided into a dry season (November–April) and a rainy season (May–October), with a marked dry spell in the month of August. The amount of annual rainfall varies between 800 and 1400 mm according to the location.

Local practice for staple crop production: In the state of Yucatan an area of approximately 170 000 ha is cultivated under a slash-and-burn bush-fallow system, locally known as *milpa*. In this rainfed system, secondary bush vegetation is cleared and burned, and the land used for maize, squash and bean production, mainly for home consumption. No manual soil preparation is done. Maize, beans and squash are sown simultaneously in the same planting hole. The period of utilization of the cleared land is reduced to a maximum of two years followed by a fallow period, which has tended to shorten over time due to the declining availability of mature forest areas. As the pressure on available land is increasing, the rotation is becoming shorter, and soil fertility, low yields and weed infestation have become serious problems.

Main characteristics of the innovative minimum tillage: This innovative practice is a combination of several technologies, including the use of cover crops, manual soil preparation and 'improved' maize varieties. Short-cycle maize varieties are sown in furrows, which are 30 to 40 cm wide. The distance between furrows is approximately 90 cm. Into these furrows the organic material produced by intercropped *Mucuna* is concentrated over the years. Every year the same furrows are used for sowing and the area is not burned any more. Farmers report maize yields that are three to four times higher than in the traditional system. Don Juan, a *campesino* who adopted the innovative technology explains the advantages as follows:

'I now have 5 mecates (one hectare contains 25 mecates) of *labranza* (minimum tillage land), and I started this work in 1994. What I harvest every year in this small area is the same amount as I get in 25 mecates of the traditional *milpa*. Apart from that, I harvest the maize in the *labranzas* one or two months before I harvest in the traditional system, and that allows me to sell fresh maize in my community during a time when nobody else has maize. The *Mucuna* helps to fertilize the soil and it also keeps the soil more humid than without *Mucuna*. This is very important during the dry spell in the summer. I have seen that my maize with *Mucuna* resists three weeks without rain. What is also very important for me is that the *Mucuna* provides forage and feed for my pigs. Therefore I prefer *Mucuna* to the other local legumes that we cultivate in the traditional *milpa*.'

Despite the increase in labour requirements, recent findings of a research study in Yucatan showed that participating *campesinos* increased the area of minimum tillage every year and nobody abandoned the technology during the four-year study period. However, the innovation did not replace the traditional *milpa* system, but rather took a niche function within the overall farming system, suggesting either that farmers are cautious of wholesale adoption of a new practice pending long-term assessment, or that they favour diversity to spread risk.

Advantages of minimum tillage:

- minimum tillage maize yields are three to four times higher per unit land
- early maize crop achieved with potential as cash crop
- soil organic matter content increases
- improved weed control
- increase in soil humidity
- additional livestock forage/feed
- permanent land-use established which does not rely on forest resources.

Disadvantages of minimum tillage:

- increased labour input for soil preparation
- *Mucuna* requires control due to its vigorous growth

> - limited availability of suitable soils for implementation
> - dependency on external seed material
> - changes and uncertainty in land tenure system (communal land use).
>
> Source: Gündel, 1998.

Lessons learned from the Yucatan case study

Lessons can be learned from this case study. First of all, the case demonstrates again the complexity of small-scale agricultural systems. The introduced innovation intervenes at different levels of the local farming system. At the field level, the cover crop has several functions (soil fertility, soil humidity, weed control) which are appreciated by both *campesino* and promoting organizations. In addition, the *campesinos* quickly discovered the benefits of the introduced cover crop for their livestock, which was a function not foreseen by the intervening organizations. The *campesinos* also discovered the market potential of an early maize crop, which further enhanced the innovation's attractiveness.

From the point of view of the external organizations involved, the establishment of permanent farming systems that are not reliant on a continuous supply of forest to maintain soil fertility, assists the process of stabilization of migratory agriculture. This, in turn, has the potential to reduce the expansion of the agricultural frontier, a goal shared by the natural resource conservation organization. However, the example from Yucatan shows that the *campesinos* do not replace their traditional system with the introduced technology, but rather prefer to combine the different activities for various reasons.

Another introduction of *Mucuna*

COSECHA has been promoting sustainable agriculture in 25 communities in Guaimaca, Honduras. Their experience with the introduction of cover crops demonstrates the advantage of offering several technologies to farmers so that they can choose among them. The main technologies promoted were live barriers and minimum and zero tillage with *Mucuna*. The aim was to increase staple food production, essentially maize and beans. *Campesinos* were especially interested in the zero-tillage technology, which shares the same characteristics as the minimum tillage described above, apart from the preparation of furrows. This technology does not require additional soil preparation. Maize is sown in rows and *Mucuna* is established between the maize rows. After the cropping cycle the organic material of the cover crop remains as a mulch on top of the soil. The zero-tillage technology is therefore less labour intensive and can be practised on larger areas. The experiences from Honduras have also shown that these innovations have led to a tripling of maize and bean yields. (Source: Gabino Lopez, COSECHA, Merida Workshop in: Anderson et al., 1997).

Due to the multiple functions of *Mucuna*-based innovations, they are able to respond to several *campesino* requirements and needs, even though the focus in promoting the technology might not be consistent with farmers' objectives.

Observations from the case studies

Cover crop species promoted in annual cropping systems
The two cases presented above both introduced *Mucuna* (spp.) as cover crops. In fact, nearly all projects presented during the Merida Workshop were promoting this species, perhaps because *Mucuna* seems to have the greatest potential to address several serious constraints on system productivity [*Mucuna* performs best in moist (over 1000 mm) and warm (over 25°C) conditions]. Farmers using *Mucuna* rotations in Mesoamerica report a wide range of benefits, including improved soil fertility, reduced weed populations and better moisture conservation (Buckles, 1994). However, in other regions *Mucuna* is largely unknown by farmers, or is not appreciated for several reasons. In many places it lacks a local market and local consumption habits, two factors that are crucial for successful adoption. In Yucatan, Mexico, farmers know a local variety of *Mucuna*, called 'Pica-Pica', which grows as a weed and contact with which causes skin irritation (also known as 'cow-itch' in Barbados). Understandably, *campesinos* were wary at first that *Mucuna pruriens* would have the same effect.

New concepts
The use of legumes within a crop arrangement is not new to farmers in Central America. Beans form an important component within staple food production. What is new to many farmers, especially those involved in slash-and-burn agriculture, is the concept of maintaining soil fertility and weed control through the management of the legumes as an alternative to burning. The promotion of a minimum-tillage system involves a series of concepts that are new to slash-and-burn farmers. First of all, it implies a radical shift from an itinerant towards a sedentary cropping system. Second, it requires labour-intensive soil preparation, which is not carried out in the traditional system. Third, it introduces a new legume species, which is frequently unknown to the farmers.

27

Function/purpose centred on soil fertility

On most occasions, the main purpose of *external* organizations in promoting cover crop technologies has been the maintenance and/or improvement of soil productivity. Their objectives are to overcome crucial problems, including soil fertility decline, organic matter losses, soil erosion, soil nutrient leaching, soil moisture reduction, and reduced water infiltration. In fact, most of the available information on cover crops looks at these issues. The impact of a cover crop technology is frequently measured in terms of yield increase of one of the main staple crops (maize, sorghum, wheat). Additional purposes of these crops, such as crop diversification, human food, animal fodder, alternative income sources and labour reduction, are recognized mainly by the farmers, but have been neglected in the past in terms of research priorities and diffusion strategies.

Focus on field level

The emphasis of external actors on soil fertility issues has led to a focus at the field level. Agronomic aspects like crop performance, yields, weed control, as well as socio-economic aspects, e.g. labour input and cost/benefit analysis, tend to be analysed and described in relation to the field where crop production takes place, not taking into account the possible interactions with other system components. The lessons from the case studies presented in this section show that farmers take a broader approach in making use of cover crops.

Perennial cropping systems

The use of cover crops in perennial systems is more widely distributed and recognized than is their use in annual crops. Indonesia is considered to be a pioneer in the use of cover crops in oil palm, coconut, rubber and sisal plantations. In fact, it has become standard plantation practice for oil palm and rubber. Cover crops in perennial plantations can be found in commercial as well as in semi-commercial and subsistence systems. Our main concern here is the role of cover crops in association with perennial species in small-scale agriculture. Nevertheless, it is important to look at commercial plantation experiences and determine how these can be adapted to smallholder situations.

Major concerns in perennial plantations are weed control, control of soil erosion, water infiltration and uniform understorey growth (leading to uniform plantation crop growth and production), which can be provided by planting one or more cover crop species. In addition to these management aspects, the provision and/or maintenance of soil nutrients is a further important consideration. In the conditions of the humid tropics, the impact of cover crops on plantation management and productivity has been positive. However, in other regions, where rainfall is not as abundant, water competition from deep-rooting cover crops has been reported (Hopkinson, 1969). Aggressive cover crops can deplete soil moisture reserves up to about 1 m depth (Lal, 1990). Therefore, the potential for cover crops is likely to be higher in regions with sufficient rainfall than for semi-arid and sub-humid regions.

Many small-scale farmers in the tropics manage reduced areas of perennial cash crops as part of their overall agricultural system. In Central America smallholder coffee, banana and coconut plantations are frequently found. In order to reduce dependency on external inputs, cover crops like *Arachis pintoi, Pueraria phaseoloides, Desmodium ovalifolium* and others are cultivated between the trees.

Functions of cover crops in different stages of perennial crop development

The functions of cover crops in perennial systems change during the development cycle of the perennial crops. During the initial phase of establishment cover crops reduce soil nutrient leaching by taking up the available nutrients that are yet not accessible to the partially developed root system of perennials. At the same time, it is important that the root systems of the two crops do not compete. In Bolivia, the association of *Arachis pintoi* and peach palm (*Bactris gasipaes*) was found to be antagonistic due to competition for nutrients, whereas the association with deeper-rooting cover crops such as *Canavalia ensiformis* appears to be satisfactory (CIAT/NRI, 1997).

Cover crops provide a quick soil cover, to protect the soil until the perennials have developed their canopy. A relevant example comes from Indonesia, where experiments have shown that soil erosion levels under tree plantations with cover crops were as low as under undisturbed forest, whereas severe erosion occurred under clean-weeded tree crops (Lal, 1990).

For smallholder farmers the establishment of a perennial plantation, whether for home consumption or commercial use, represents a major expenditure and a potential loss of productive area. It is important to seek ways of reducing establishment costs, and of providing short- to medium-term income and/or food from the plot until harvest of the perennial crop commences. In Bolivia the principle of sequences has been incorporated into the design of novel farming systems, as shown in Table 4. It can clearly be seen how the food and/or cash provision from the plot shifts from annual to semi-perennial to perennial crops over time.

Table 4 Sequences of crops/cover crops during the establishment of smallholder plantations in lowland Bolivia

First summer	First winter	Second summer	Second winter	Third summer	Third winter
Plant citrus	Citrus development phase	Citrus development phase	Citrus development phase	Citrus development phase	Citrus start production CASH
Plant pineapple	Pineapple development phase	Pineapple start production CASH	Pineapple in production CASH	Pineapple in production CASH	Pineapple in production CASH
Rice, cassava or maize between perennials FOOD/ CASH	Beans, cowpea, groundnut or cover crop between perennials FOOD/ CASH	Rice, cassava or maize between perennials FOOD/ CASH	Beans, cowpea, groundnut or cover crop between perennials FOOD/ CASH	Cover crop (perennial or annual)	Cover crop (perennial or annual)

In this example, farmers can substitute other perennial crops for citrus (e.g. macadamia, tamarind, etc.) and other semi-perennials for pineapple (e.g. bananas or paw-paw). The intercropped annual food crops during winter and summer can vary according to family needs and preferences, and the cover crop species can also be chosen from a range of annual or perennial species according to the management regime the farmer wishes to apply. Perennial covers such as *Arachis pintoi* and *Pueraria phaseoloides* are slower to establish (see Figure 1), but should then require relatively low maintenance, whereas annual covers allow more flexible management in response to needs and conditions.

In lowland Bolivia, where land is relatively abundant but labour and capital are limited, it is very important for smallholders to be able to reduce labour costs and generate income quickly. The use of cover crops and sequences in flexible combinations provides these requirements, giving a staggered cash and food supply over time (adapted from Ardaya et al., 1998).

Tracing back through the history of *Mucuna* in Central America one also arrives at a plantation system. *Mucuna*, known and commercialized in the United States as 'banana field bean', was probably introduced by the United Fruit Company in the banana plantations along the Atlantic coast of Central America. Its main purpose was to provide fodder for the mules that were used to transport bananas from the plantations to the depots (Buckles, 1994).

Other common cover crop species intercropped in perennial plantations are *Pueraria phaseoloides, Desmodium ovalifolium, Arachis pintoi* and *Calapogonium mucunoides*. In Costa Rica, for example, cover crops are integrated into guanabana (*Annona muricata*) plantations, while in Honduras and Surinam *Mucuna* is established in many citrus plantations, and in Panama cover crops can be found in banana plantations (CIDICCO 1994). Table 5 highlights the experiences of the Merida Workshop participants regarding the use of cover crops in citrus and coffee.

Table 5 Cover crop uses in perennial crops

Perennial crop	Cover crop	Positive aspects	Negative aspects	Country, Location
Citrus	1. *Canavalia ensiformis*	Produces lots of organic matter; rapid establishment and good cover; good adaptation range, grows well under shade; does not climb; few pests	Difficult to collect seeds due to sequential maturity; not good as animal feed	CIAT, Bolivia, 400 m above sea level, 1200–1800 mm rainfall
	2. *Cajanus cajan*	Edible, green manure	Dries quickly, limited weed control	CENTA, El Salvador
	3. Grass	Controls erosion; weed control; is not host for pests	Competition for nutrients. Need to cut around trees	CIAT, Bolivia
	4. *Mucuna*	Contributes to soil fertility; good weed control; controls *Imperata*; potential animal feed	Eaten by sheep; needs replanting each year; does not grow well under shade; establishes slowly in dry areas; some attack by soil pests	Honduras CIAT, Bolivia CENTA, El Salvador

Coffee	1. *Arachis pintoi*	Weed control good once established	Slow establishment	Honduras, 600 m above sea level; 1400–3000 mm rainfall
	2. *Canavalia* spp	Good weed control; does not climb over coffee; Perennial (?); lengthens the harvest period; grows well under shade	Leaves die in cold temperatures; is not dual purpose (i.e. cannot be eaten)	Honduras
	3. *Tephrosia*	Good shade; sufficient organic matter; weed control	No edible seeds	
	4. *Cajanus cajan*	Aids soil fertility; grows fast; virus barrier; human food; allows enough light to plant other crops	Lasts only two years; cannot be used in mature plantations	Mexico, 800–1200 m above sea level, 1400 mm rainfall

Case study of farmers' perceptions of cover crops in perennials

The case study presented below is from lowland Boliva, and looks at farmers' perceptions of different cover crop combinations in perennial crops. The case study was presented by CIAT/NRI during the cover crops workshop in Merida (Anderson et al., 1997).

Box 12 Citrus and pineapple in association with different cover crops

The CIAT project (Adaptive Research in Sara and Ichilo) is supported by two projects managed by NRI and funded by the UK Department for International Development. The aim of the project is the development of sustainable resource-use strategies at the humid forest frontier. The research approach taken by CIAT/NRI is presented in Chapter 6. As part of this project in the tropical eastern lowlands of Bolivia, cover crop use in citrus/pineapple inter-crop systems has been promoted. A study by Warren (1997) looked at farmers' perceptions of this innovative practice. The study was carried out in five of the 89 communities participating in the project. Before the project started,

---

the farmers were not aware of the potential benefits of cover crops within their perennial crops but they were interested in trying out the proposed innovation. Four different cover crop species were given to the farmers. These species were *Arachis pintoi*, *Mucuna pruriens*, *Canavalia ensiformis* and *Cajanus cajan*.

The positive and negative aspects perceived by the participants for each species were the following:

Arachis pintoi:

☺ Positive aspects	☹ Negative aspects
good weed control (once fully established), good soil cover, easy to manage	slow establishment, insufficient soil cover, needs a lot of labour input, not suitable for larger areas, invades pineapple plant, difficult to eliminate

General conclusions of the farmers were that they did not like *Arachis* very much because it requires high labour input and does not control weeds sufficiently. Most of them decided, therefore, to substitute *Arachis* with a different cover crop species.

Mucuna spp.:

☺ Positive aspects	☹ Negative aspects
controls weeds, fertilizes soil, easy to manage, climbs up the pineapple and provides shade for the fruit	climbs up the plant, reducing flowering, eaten by sheep

General conclusion: Farmers like *Mucuna* because of the positive impact on weeds and they want to continue using *Mucuna*.

Canavalia ensiformis:

☺ Positive aspects	☹ Negative aspects
fertilizes soil, controls weeds, not eaten by sheep, maintains soil moisture, good soil cover, good development of other crops, quick and strong growth	no seed production because of high rainfall (exceptional circumstances), no market for seeds (market situation is highly variable depending on locations), shades other plants and has to be cut

General conclusion: There is great interest from farmers in using *Canavalia* because of successful weed control, and the fact that *Canavalia* can be easily managed.

Cajanus cajan:

☺ Positive aspects	☹ Negative aspects
fertilizes the soil, good for human consumption, good rice harvest after *Cajanus*	dries up quickly, insufficient weed control

General conclusions: If *Cajanus cajan* is densely established (reduced plant distances) then it controls weeds successfully; apart from that, farmers like it because the beans are edible by humans.

Farmers also tested *Mucuna pruriens* and *Calopogonium mucunoides* as cover crops within banana plantations. *Mucuna* was the preferred cover, controlling weeds well (including *Imperata contracta*). In addition, Bolivian farmers tried a number of covers (*Arachis pintoi*, *Canavalia ensiformis*, *Flemingia macrophylla*, *Pueraria phaseloides* and *Cajanus cajan*) in association with peach palm (*Bactris gasipaes*). Briefly, their conclusions were that *Arachis pintoi* competes with the peach palm and is therefore unsuitable, *Pueraria phaseoloides* is too aggressive, and *Cajanus cajan* does not control weeds sufficiently. Their preferred option is *Canavalia ensiformis*, even though it is not aggressive enough to control *Imperata contracta*.

The criteria used by farmers to evaluate the integration of cover crops into perennial system are shown in Figure 2.

Other experiences of farmers with cover crops in perennials

More information on the use of cover crops in perennial plantations is provided by CIDICCO, who have analysed an experience in Honduras where *Mucuna* has been intercropped with citrus trees. Box 13 summarizes this experience:

Figure 2 Farmers' criteria used to evaluate the different cover crops

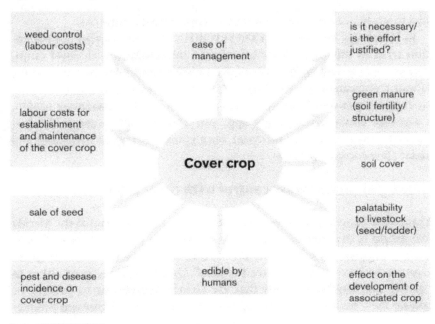

Source: CIAT/NRI, 1997

Box 13 The use of *Mucuna* in citrus plantations: The experience of Zopilotepe, Honduras

An abandoned citrus plantation, which was rehabilitated a year ago, was used as a demonstration plot to show farmers the positive impact of *Mucuna* on poor soil quality and weed infestation. Between the citrus trees, which were planted at a distance of 4 m by 4 m, three rows of *Mucuna* were sown with a spacing of 1 m between plants. [NB The planting distances are location-specific. In Bolivia for instance, a spacing of 40 by 40 cm was necessary to provide rapid and complete cover with little or no weeding.]

The *Mucuna* was cut back frequently after an establishment period of 40 days. An observation of the technicians in charge: 'After four months the yellowish-green colour of the citrus tree foliage changed into a dark green colour. The trees developed new sprouts and they appeared to be healthy.' Once the *Mucuna* is established it reaches a 100 per cent ground cover and weeding is not required anymore. It was estimated that 30–50 t/ha green organic material was produced by the *Mucuna*.

Source: CIDICCO, 1992

Other experiences of cover crop uses in forest plantations are reported. However, the information provided during the Merida Workshop was limited to studies carried out so far under experimental conditions.

In Honduras, for instance, CONSEFORH (1997) has carried out evaluation trials of different timber tree species intercropped with cover crops. The aim was to reduce the costs of plantation management by reducing the labour requirements for weeding. A further aim was to improve soil fertility to enhance tree development. Cover crop species like *Neonotonia wightii, Canavalia ensiformis, Mucuna* spp. and *Dolichos* spp. were established beneath pine trees (*Pinus caribaea*), eucalyptus (*Eucalyptus citriodora*) and cedar (*Bombacopsis quinata*).

Weed, pest and disease control with cover crops

One major function of cover crops cited in the literature and at the Merida Workshop is the control of weeds in annual and perennial crops. Bunch (1997) suggests that 'perhaps the second most common use of green manure and cover crops is weed control'. Cover crops are used either to control and eliminate weeds that are already present, or to prevent or reduce their establishment. Initial aggressive growth and biomass production (e.g. *Mucuna*) compared with weeds can lead to the smothering and elimination of established weeds. Examples are the control of *Imperata cylindrica* in Indonesia (Gurtino et al.) and Benin (Versteeg and Koudokpon, 1990). A Latin American example is presented below – from Bolivia, where the grass weed *Imperata contracta* is troublesome in perennial orchards in the tropical lowlands.

Box 14 *Imperata contracta* **control in Bolivia**

Three cover crops were evaluated by CIAT/NRI (reported at the Merida Workshop) for their efficacy in controlling *Imperata contracta*, which is a serious weed in the transition from slash-and-burn farming to more settled and intensified systems. Table 6 presents data on the survival of *Imperata* rhizomes after 20 months of cover by the three covers, concluding that the *Mucuna pruriens*, known locally as 'grey Mucuna' (*Mucuna pruriens*, subspecies cinereum) is the most effective at controlling the weed. For most effective control the cover should be planted very early in the rainy season, so as to cover the ground before the *Imperata* has emerged.

Table 6 Survival of *Imperata* rhizomes under cover crops in Bolivia

Treatment	Number of live rhizomes	Number of dead rhizomes
Without cover	99	14.5
Mucuna pruriens (cinereum)	0	66.2
Mucuna pruriens (aterrinum)	4.2	42
Calopogonium mucunoides	2.8	23.2

(sampled 20 months after establishment of covers, to a soil depth of 25 cm)

Source: Anderson et al., 1997

The shade and soil cover (either green or dead) provided by cover crops reduces the penetration of ultraviolet radiation and decreases soil temperature, which leads to reduced germination of weed seeds. Table 7 shows a ranking exercise done by a group of 20 farmers in Yucatan, southern Mexico. They compared seven legumes in terms of their capacity to control weeds and to establish ground cover. The exercise shows how farmers took possession of the concept of cover crops for weed control and how they applied the new concept to evaluate local legume species.

Table 7 Suitability of different legumes for weed control and ground cover

Legume	Weed control	Ground cover
Beans – *Phaseolus vulgaris*	limited impact	good
Lentils – *Lens culinaris*	no impact	reduced
Xpelon – *Phaseolus lunatus*	good	very good
Jicama – *Pachyrhizus erosus*	good	very good
Ibes – *Vigna* spp.	good	good
Mucuna – *Mucuna pruriens*	very good impact	very good
Canavalia – *Canavalia ensiformis*	good	good

Allelopathy can reduce the germination and development of other crop species. This effect occurs through the production of biologically active plant substances which interfere directly or indirectly with other, sensitive plants. An example of a cover crop with allelopathic effects is *Mucuna pruriens*, which has shown a significant ability to reduce both the number and biomass of weeds when planted in rotation with maize (Gliessman in Rizvi and Rizvi, 1992). The concentration of allelochemicals varies according to the phenological stages of *Mucuna*. Escarzega (in Rizvi and Rizvi, 1992) found that dicotyledonous weeds are more inhibited by leaf leachates of flowering plants, and monocotyledonous weeds by plants in fructification.

An experiment carried out by CIAT in the Santa Cruz Department, Bolivia, compared two cover crops (*Mucuna pruriens* and *Arachis pintoi*) in a citrus–pineapple plantation with management without cover crops. Table 8 presents the results obtained. *Mucuna* proved to be most effective in terms of weed control. The good weed control by Mucuna depended on the rapid establishment of full cover, which can be achieved through early planting within the rainy season, good density of population (in Bolivia 40 cm x 40 cm), and good germination (seed quality, soil moisture and date of planting).

Table 8 Comparison of different legume species for weed control

Cover crop	*Mucuna*	*Arachis pintoi*	Without cover crops
Type of weed	Weed biomass g/m^2	Weed biomass g/m^2	Weed biomass g/m^2
Dicotyledons	0	6.7	196
Gramineae	0	322	407
Cyperaceae	0	35	5.5
Shoot regrowth of bush species	62.5	31.9	101
TOTAL	62.5	395.6	709.5

Source: Merida Workshop; Anderson et al., 1997

In some situations cover crops themselves are potential weeds. Farmers are concerned about the elimination of cover crops from their fields once they have decided not to cultivate them any more, or before planting an annual crop whose development might be affected by competition from the cover crop (e.g. rice).

Pest and disease control

The literature indicates a positive impact of high levels of organic matter on the incidence and severity of root diseases. The control of nematodes by cover crops has been tested in various studies. *Cajanus cajan* and *Mucuna pruriens* showed good potential for the prevention of reproduction of different nematodes (Haroon and Abadir, 1989). The transmission of soil-borne diseases to other parts of the crops is reduced when cover crops are used to reduce rain splash – an important natural agent for the dispersal of diseases.

Cover crops such as *Calopogonium caeruleum* and *Paspalum conjugatum* were found to reduce the intensity of attack and viability of the rubber tree pathogen *Rigidoporus lignosus*. However, these crops were unable to prevent the spread of the disease once the collar of the tree had been infected (Sinulingga et al., 1989). Cover crops have also been used in crop rotations to reduce the population of pathogens in the soil (Gold and Wightman, 1991).

On the other hand, cover crops might contribute to the increase of pest and diseases, especially if they serve as host plants. Different pests and diseases attack cover crops, and it is important to not to associate such a cover crop with other crops susceptible to the same pest or disease.

Farmers report increases in snake, rodent and insect populations in their fields, where these animals find a protective cover in the cover crops. In regions where snakes present a serious problem, farmers tend to cut down the cover crops before entering their fields for harvest. Farmers are also reluctant to stop burning, as this has traditionally been used as a pest control device.

There is a need to carry out more integral research on the impact of cover crops on pests and diseases, as this offers potential for sound integrated pest management measures for resource-poor farmers (Thurston and Abawi, 1997).

Conclusions

The experiences with cover crops in annual and perennial crops presented here, as well as the results of the discussions during the workshop, have highlighted a number of issues. These are summarized below.

Integration of cover crops in perennials with other system components
The diversification of perennial crops with cover crops, annuals and semi-perennials and/or the integration of animals is an important strategy to recoup the capital investment associated with perennial crop establishment and obtain short-, medium- and longer-term benefits from the land. There is a lack of information and expertise perceived by development organizations

on the integration of animals in perennial cropping systems, especially on the use of seeds and foliage for animal nutrition. Chapter 3 provides information on this issue, and presents the potential for cover crop/livestock integration.

Impact of cover crops on the economic and agronomic performance of perennials

The experiences stress weed control (and hence the reduction of maintenance costs) as being an important function of cover crops in perennials. Information is scarce on the impact of cover crops on main crop yields and crop development, and the impact on pests and diseases. The experiences of the use of cover crops in commercial plantations should be documented to assist those interested in the selection of suitable cover crop species and management practices for smallholder situations. Good understanding of the functions and purposes that cover crops might have in smallholder perennial production is required to increase the relevance of research and dissemination, and thereby the adoption of these technologies.

A general lack of communication and systematization of existing experiences limits a learning process between formal organizations (research and development organizations) and farmers.

Selection of cover crop species and varieties

Farmers' circumstances have been shown to be very varied between and within the case studies presented in this chapter. A similarly wide choice of cover crop species and varieties is required so that farmers can choose those that fulfil the purposes and functions they require, and perhaps additional purposes felt to be desirable by external organizations. Making this wide choice available will require careful and systematic identification and documentation of local cover crops and their uses, and the classification of local and exotic species in such a way that enables their potential application to a particular set of conditions to be recognized. The Bolivian case has shown that farmers have a long list of criteria by which to evaluate different species and their performance in specific crop combinations.

The availability of cover crop species for farmers and development organizations can be an important constraining factor to widespread adoption. Increasing cover crop seed demand requires the establishment of multiplication and distribution mechanisms (both formal and informal).

Food, feed and forage

Our concern in this chapter is to explore the benefits and drawbacks of different forms of cover crop/livestock integration. We shall use the terms 'food', 'feed' and 'forage' to denote the products and by-products of crops, pastures, cover crops and livestock. Crop, cover crop and livestock products consumed by people are 'food'; crop and cover crop grains, tubers and non-leaf products that livestock consume are 'feed'; and other crop, pasture and cover crop products are 'forage.' Products are the outputs from components of an agricultural system which the farmer considers the main purpose of managing the component. By-products are a component's secondary outputs from the farmer's point of view.

Before considering the ways in which cover crops and livestock can be integrated we shall discuss the human food aspects of cover crops, especially legumes.

Legumes as part of the human diet

Legumes are an important part of people's diet in different regions of the world – especially in poorer regions, where they often provide the only regular source of proteins, or amino acids. The consumption of legumes is variable. In Latin America relatively high levels are consumed, 40–70 g dry matter (DM)/day/person, compared with Sri Lanka, Malaysia and Indonesia with 15–20 g DM/day/person (Borget, 1992). Apart from their high protein content, legume seeds contain appreciable levels of carbohydrates and some are also rich in oils. Table 9 presents the nutritional value of some legume species common in the tropics.

LEIA farmers generally consider the edibility attributes of legumes used as cover crops as important. In their traditional agricultural practices the significance of the multiple uses of cover crops is well known. In Chapter 2 we presented some examples of traditional cover crop associations (Table 3). In small-scale agricultural systems legumes are commonly

Table 9 Nutritional content of some legume species common in the tropics*

Legumes	Proteins	Fats/oils	Carbohydrate
Groundnut	20–33	42–8	22–5
Pigeon pea	15–29	1–3	60–6
Soya bean	37–41	18–21	30–40
Dolichos lab-lab	24–8	1–2	65–70
Lima bean	19–25	1–2	70–5
Common bean	20–7	1–2	60–5
Winged bean	30–40	15–20	35–45
Cowpea	22–6	1–2	60–5

(* contents g/100 g of entire mature seed) Source: Borget (1992)

cultivated with other staple crops like maize and millet, as intercrops, relay crops or in rotations (e.g. in Central America: maize and *Vigna* spp. and/or *Phaseolus* spp.).

The contribution of cover crops to human diets has been neglected in many projects and studies. The case study by CIDICCO in Chapter 2 shows how the edibility of cover crops is a prime concern of farmers and leads to preferences between species. Introduced cover crops have been chosen, largely by scientists, on the basis of characteristics other than their attributes as components of human diets. However, farmers immediately raise this question when they are introduced to a new species. *Mucuna* spp. is considered to be an edible legume in Java and Nepal, where it is traditionally prepared as a fermented beverage. In Central America efforts have been made to promote the use of *Mucuna* as a constituent of beverages, tortillas and bread. However, these promotion efforts have been hindered by people's wariness of a toxic component (L-dopa) in *Mucuna* grain, which requires it to be cooked before it is safe for human consumption. Cases of dizziness, vomiting and short-term effects on eyesight have been reported after the consumption of *Mucuna*. Recommendations on how to avoid the negative effects of *Mucuna* consumption include boiling, changing the water and limiting the amount of *Mucuna* intake. In situations where farmers already produce local legume species for consumption it might not be necessary to promote the use of *Mucuna* for human diets as it can make a major contribution to the animal diets (see evidence from recent Mexican research, below.)

Figure 3 shows the different possible contributions of cover crops as food, feed and/or forage.

Figure 3 Possible uses of legume cover crops for food, feed and/or forage

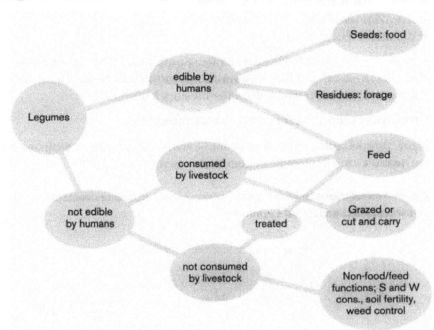

Non-legume plants that can be used as cover crops, such as sweet potato (aerial parts of plant used as animal forage) and cucurbits, also provide food. There is a real need to seek out further cover crops with a human food potential.

Livestock within low external input agriculture

More than 85 per cent of all livestock (ruminants, pigs and poultry) that exist in the tropics are in small-scale agricultural systems (Wilson, 1995). The purposes of commercial livestock producers are well known – profitability through the sale of meat, milk, eggs, skins, wool and fibres, etc. The purposes that semi-commercial and subsistence producers – those largely involved in LEIA – have in owning livestock we understand less well. Wilson (1995) presents a list of purposes that LEIA producers might have in owning livestock:

- to diversify production in order to reduce risk
- to generate and accumulate capital
- to provide services for crop production (work, fertilizer, fuel)

- to fulfil customs and rituals
- to lend status and prestige to the owner
- to provide food and other products
- to generate income.

Not all LEIA producers will share all these purposes. However, we can conclude that livestock in LEIA have various functions and farmers have different multiple purposes which drive their livestock husbandry.

Crop, cover crop, pasture and livestock interactions

Figure 4 shows some of the interactions between crops, cover crops, pasture and livestock in an integrated agricultural system.

Figure 4 Interactions between crops, cover crops, pasture and livestock in an integrated agricultural system

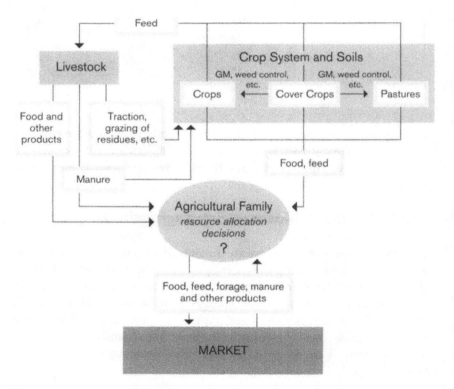

From the figure we can see that crops produce food for the agricultural family and for sale, and feed and forage for livestock; cover crops fulfil green manure and weed control functions for soils and crops, and produce feed and forage for livestock. They might also produce food for the agricultural family; livestock produce food and other products for the agricultural family and for sale, and manure for the soils and crop systems. When livestock can be used for traction, cover crops can be ploughed into the soil or crushed and chopped using a bladed roller prior to planting a staple food crop, as occurs in Brazil.

We can appreciate the number of trade-off decisions that the agricultural family has to make in the allocation of products and by-products produced in the agricultural system and utilized by other components, or sold to the market. Table 10 lists some of these. The trade-offs of most interest here are those where cover crop products and by-products are allocated to soil and water conservation, soil fertility and soil cover roles – which we have called green manure and weed control functions, and for feeding livestock.

Livestock contribute to crop production both in direct ways – work, manure, etc. – and in indirect ways. Livestock products are quite often of high value, and can usually be sold easily at times convenient to the vendor. Hence, income from livestock sales can be invested in crop production (including cover crops) in a timely way. By converting crop and/or cover crop products and by-products into livestock products, economic value is often aggregated. For instance, by feeding maize stover to small ruminants, or legume grain to pigs or poultry, the income generated from the sale of the livestock product, allowing for the conversion efficiency, is greater than that which would have been achieved by selling the crop by-product.

LEIA producers seek synergies between the components of their agricultural systems. In this respect the utilization and cycling of nutrients are critical factors in the husbandry of crops, cover crops and livestock. In mixed agricultural systems nutrient cycling is often mediated through livestock. Livestock increase the farmer's options by which s/he manages nutrients in terms of their spatial and temporal placing. They can also be used to accelerate nutrient turnover in the production cycle, synchronize nutrient demand and supply, and reduce nutrient losses (Steinfeld et al., 1997). Incorporating legumes into an agricultural system is one of the very few ways to generate new nutrients (rather than importation as fertilizers, composts, silt, etc.). The harnessing of nutrient generation with the enhanced capacity for nutrient allocation, recycling and maintenance through crop/livestock integration demonstrates some of the possible synergies that can accrue through combining cover crop and livestock husbandry. Stobbs (1969) demonstrated how including legumes in pastures not only improves animal performance but also has positive effects on the yields of crops that follow the pasture/legume association.

Table 10 Some of the trade-offs involved in crop/pasture/ cover crop/livestock integration

Resource	Trade-off decision	Possible outcomes	Comments
Land	Allocation between market, crops, cover crops, pastures or fallow	1 sold or rented 2 sown to crops 3 sown to cover crops 4 sown to pastures 5 left fallow	2 to 5 All outcomes can be combined temporally or spatially
Crop– grain	Allocation between family, farm, market or livestock	6 family consumption 7 sold 8 seed for crops 9 feed for livestock	6 to 9 Outcomes mutually exclusive but available grain can be proportioned to different outcomes
Crop– forage	Allocation between farm, market or livestock	10 organic matter for composting or soil incorporation 11 forage for sale 12 forage for livestock	12 Forage transformed in to livestock products (with possibility of aggregated economic value) including manure. Forage can be cut and carried to livestock and manure collected, or livestock folded over crop land
Cover crop– grain	Allocation between family, farm, market or livestock	13 family consumption 14 sold 15 seed for cover crops 16 feed for livestock	13 and 16 Anti-nutritional factors may be a problem.
Cover crops– forage	Allocation between farm, market or livestock	17 green manure 18 forage for sale 19 forage for livestock– grazing or cut and carry	17 Cover crops can be used to regenerate degraded pastures 19 Direct competition between weed control and green manure roles and utilization for livestock. However, completely under farmer's control. Cutting might stimulate regrowth, roots not affected by grazing.
Livestock– food and other products	Allocation between family and market	20 draught and management 21 family consumption 22 sold	

Livestock–manure	Allocation between family, market and farm	23 fuel for family 24 sold 25 fertilizer	24 Timeliness and precision in spatial allocation; way of transferring nutrients from low to high potential, or to areas of greatest need.

We should also recognize that there are a series of limitations in the use of cover crops for livestock:

- The price and availability of cover crop seed varies tremendously. Limited supply results in high prices and hence livestock feeding is not viable. Low prices often act as a disincentive to harvesting the seeds; hence, in the short term until price responds to demand, availability is adversely affected.
- Grazing of cover crops might adversely affect the development of the plant and restrict the crop's ability to fulfil other biophysical functions (in other cases grazing can be used to reduce crop/cover crop competition).
- Husbandry practices that can optimize the utilization of cover crops for various functions, including green manuring, weed control, soil moisture conservation and providing livestock feed and forage, need to be investigated .
- Anti-nutritional factors are found in many cover crops that inhibit their use for livestock feeding, and resources have to be allocated to their treatment prior to feeding.

Table 11 shows some uses of cover crops for livestock feeding found in Central America. Before the cover crop grains such as *Dolichos lab-lab*, *Mucuna* and *Canavalia* are fed to livestock, treatments are carried out to reduce anti-nutritional factors and to improve voluntary intake. These treatments include toasting, boiling (changing the water periodically), grinding, soaking and germinating (see below for results of work on feeding *Mucuna* beans to chickens).

Table 11 Cover crops used for livestock feeding in Central America

Country	Cover crop and use
Mexico	White clover fed to sheep and cattle (Mixteca, Oaxaca), Maize stover and *Canavalia* beans fed to goats (Yucatan), *Mucuna* bean and forage fed to pigs (Yucatan)

El Salvador	Sorghum straw and *Dolichos lab-lab* fed to cattle,
	Forage peanut fed to cattle and goats
Honduras	Maize stover and *Dolichos lab-lab* fed to cattle and horses,
	Campanilla (*Vigna*) fed to cattle and horses

Source: Anderson et al., 1997

During the Merida workshop on cover crops (Anderson et al., 1997) a working group consisting of representatives of NGOs, farmers' groups and researchers discussed the integration of cover crops and livestock in mixed agricultural systems. Their conclusions on the socio-economic aspects are presented in Box 15.

Box 15 Conclusions from discussions on the socio-economic aspects of cover crop/livestock integration

It is considered that socio-economic aspects of cover crop/livestock integration are less well understood than technical issues. However, the following points are important:

- the hypothesis that cover crop use can increase staple food production and reduce the amount of staple grain fed to livestock requires testing under different conditions
- the integration of livestock into an agricultural system depends upon the productive capacity of the farm and the resources available to feed livestock. It is important to identify carefully the families with the capacity and interest to incorporate livestock into their agricultural system before any such project is initiated, and in order to do so a participatory appraisal is required
- without an adequate market for livestock products, the potential for integrating livestock and cover crops into an agricultural system to provide for seasonal shortages of food, and/or an extra income source, will not be achieved and, worse still, the need to feed the livestock could prove a severe burden for the family
- the integration of livestock can result in increased demand for rented pasture and/or purchased feed or forage (from cover crops) thus stimulating the local market, and transactions between families with access to different resources
- at regional and national level, feeding livestock with cover crop products may provide an alternative to imported grains and concentrate feeds

---

- livestock husbandry has the potential to improve family diets. However, the promotion of feeding grain to livestock (notably poultry and pigs) should not take place unless family food security is assured. The introduction of cover crops suitable only for livestock and not human consumption should take place only in circumstances where the achievement of family food security is possible using other resources
- the use of cover crops for livestock feeding could impose significant increases in the workload of female family members, especially where treatment of the grain is necessary. The same treatment might also mean an increased use of resources such as firewood and water, which might be scarce.

Source: Anderson et al., 1997

Impact on the environment

Mixed farming is considered beneficial for the land when soil fertility is maintained by crop rotations, and the incorporation of legumes (cover crops) that function as nutrient providers and forage sources and which restrict erosion. Integrating livestock into an agricultural system and incorporating manure does not generate nutrients or reduce nutrient surplus *per se*. What this form of husbandry can do, if so managed, is to increase soil organic matter and enhance soil micro-flora and fauna.

The key question is one of maintaining the nutrient balance, and here we have to be very clear about what size of land unit we are considering. For example, depleting nutrient reserves from land around the farm by putting animals on to common grazings during the day, and then collecting their manure (from night pens) for use on the farm, might maintain the environment of the farm while damaging the wider environment. Hence, innovations that allow the balancing of nutrient status within the farm help to maintain the wider environment. Crop/livestock integration offers management tools to the farmer that assist in the balancing of farm nutrient status. The inclusion of leguminous cover crops in the mixed farm provides the farmer with a broader repertoire of management options and, most importantly, the potential for generating nutrients from available resources (it is often assumed that legumes have active rhizobia; this may not be the case, particularly for immature plants).

Conservation NGOs, particularly those working in the buffer zones around protected areas, are looking for ways of influencing farmers' practices so that the negative impacts of land husbandry are reduced and positive impacts enhanced. A case in point is the work of Pronatura Peninsula Yucatan in south-eastern Mexico which, together with the Autonomous University of Yucatan, has promoted cover crop use by *campesino* farmers in the buffer zone around the Calakmul Biosphere Reserve. Box 16 describes this project. The NGO decided to develop homegarden pig-keeping as a way of consolidating the use of *Mucuna* as a cover crop in maize production. Studies are under way to evaluate the impact of this initiative. However, the basis of the project is the enhancement of nutrient cycling in the maize plot/forest/homegarden system. This has been attempted first through the introduction of cover crops to improve soil fertility and reduce weed invasion, and hence extend the periods of maize cultivation on the same plot; and, second, the introduction of *Mucuna* feed to pigs to aggregate the economic value of the *Mucuna* bean and thereby encourage farmers to cultivate *Mucuna*.

Box 16 Maize, *Mucuna* and pigs in Calakmul, Mexico

The Calakmul Biosphere is part of a tropical forest that extends from the Petén in Guatemala, into north Belize, across the Mexican states of Quintana Roo and Campeche, and into Chiapas. It is the most important tropical forest of the Americas north of the equator. Unfortunately the forest is not intact, being formed of separate reserves in the three countries mentioned. The Biosphere is made up of 227 860 ha of nucleus and 494 140 ha of buffer zone. There are 72 communities within the buffer zone, the majority of whom have been displaced from other regions of Mexico.

Pronatura Peninsula Yucatan is just one of the development organizations working in the zone. Their mission is to contribute to the conservation of ecosystems of the Yucatan. In Calakmul, Pronatura are involved in projects to develop organic agriculture, beekeeping, homegardens, reproductive health, and eco-tourism. Maize is the staple food of all the Calakmul communities and it is produced through a slash-and-burn agriculture. These practices are considered to threaten the biosphere and Pronatura has sought to ameliorate the effects of maize production by introducing cover crop use and other forms of organic agriculture.

After three years of pushing *Mucuna* use in the Calakmul region, concern was felt due to the low adoption rates and abandonment of cover crop technology by farmers, and it was decided to develop other uses for cover crop products. In association with the Autonomous University of Yucatan, Pronatura embarked upon a project to promote the use of *Mucuna* beans as feed for small-scale pig production. The objective of the project was to develop a system of livestock production integrated into the organic crop system in such a way that efficient use can be made of crop products leading to an aggregated economic value.

Farming families that had sown *Mucuna* were invited to participate in the project. They were given one or two sows, on a revolving fund basis, once they had built suitable pig pens and had attended workshops where pig husbandry knowledge was discussed.

The project has been running for some years and initial results are available on sow and litter performance (see later in the chapter). The impact of the project on *campesino* agricultural systems is at present being evaluated.

Source: Case study presented at the Merida Workshop on Cover Crops by Pronatura Peninsula Yucatan, in: Anderson et al., 1997.

Little scientific information exists about the environmental and ecological impacts of cover crop/livestock integration in agricultural systems. While acknowledging this dearth, the participants in the regional workshop on cover crops (Anderson et al., 1997) concluded that:

- the integration of cover crops and livestock can improve the bio-economic efficiency, and hence the stability, of an agricultural system both in spatial and temporal dimensions. In turn, this improvement in stability could have a positive effect on the conservation of natural resources both within and around the farm.
- livestock husbandry could reduce the need to hunt wild animal species to augment the family diet. The time saved by not hunting could then be usefully employed in farm management.
- the integration of cover crops and livestock is a means of intensifying the management of the agricultural system, enabling the support of a higher human population per unit area. In turn, this intensification of land use would reduce the need to exploit the remaining natural resources (including forests) and biodiversity.

Despite these potential advantages, there are relatively few documented examples of cover crops being used successfully as livestock feed or fodder. Farmers and researchers are working together in south-eastern Mexico in an attempt to unleash this potential.

Evaluating cover crops as livestock feed

A programme of applied research has been initiated at the Autonomous University of Yucatan to identify the feeding values of crops and by-products identified in the rural appraisals of *campesino* agricultural systems, and to test appropriate methods for reducing anti-nutritional factors in legumes identified as potential sources of livestock feed and forage. This is part of a collaborative research project between Wye College, University of London, and the Vet Faculty of the Autonomous University of Yucatan, 'Optimising the integration of livestock into low external input crop systems', sponsored by the Livestock Production Programme, DFID, UK. The objective is to produce results useful for application in *campesino* systems for livestock feeding. In order to achieve this objective the researchers have had to adjust themselves to the characteristics and needs of the the *campesino* communities, most notably in terms of the crops and by-products and the species of livestock considered. As part of the same project, pig husbandry in *campesino* production systems has been monitored. Diets have been recorded and animal performance measured.

The sequence of trials to identify the feeding values of crops and by-products is:

- short-term tests of voluntary intake (diets with different inclusion levels)
- animal performance tests (the same diets over longer periods)
- digestibility trials.

The methodological issues considered are:

- the control diets used in the experiments have to be appropriate points of comparison for *campesino* systems
- the material is tested taking into account the possible forms of use under different conditions, for example: a) where the availability of the livestock base diet (usually maize grain) is sufficient during the whole year, and so the tested material is to provide a supplement to the diet that will improve its quality, and b) where the material tested is to replace the base diet due to scarecity
- the treatments tested to improve consumption should minimize the use of resources such as time, energy, and water.

Taking account of these considerations, the applied research does not seek to find the ideal diet nor a unique solution. The intention is to respond to the real needs appraised with the *campesino* communities. In Table 12 the analytical framework for testing a protein-rich source is shown.

The rural appraisals have shown the importance of poultry and pig husbandry to the livelihood strategies of the families in marginalized *campesino* communities. Innovations in *campesino* crop systems include the use of cover crops, especially *Mucuna* and *Canavalia*. The opportunity to use cover crop products for livestock feeding was identified.

The following tables present a summary of the results achieved so far. Tables 13, 14, 15 and 16 present data from the evaluation of *Mucuna* and *Canavalia* as feed for poultry. All the diets, including the controls, are iso-proteinic and iso-calorific. Table 17 presents observations of pig performance from small-scale husbandry systems in the communities of the Calakmul Biosphere buffer zone.

Table 12 The analytical framework for the evaluation of *Mucuna* grain as a feed for monogastric livestock in *campesino* systems

Availability of feed fractions in the system	Purpose of the use of *Mucuna*	Consumption level of the *Mucuna* required	Is *Mucuna* treatment necessary?
Energy sufficent, protein sources very limited	Protein supplement	To provide sufficient protein	NO – voluntary consumption of raw grain enough
Both energy and protein limiting	Main component of the diet	To cover the main part of nutritional requirements	YES – to improve intake levels, but without reducing nutritive value

Table 13 Short-term intake by chicks of diets containing *Mucuna* or *Canavalia* grain

Treatment* (% inclusion level)	H O U R S							
	0.5	1	1.5	2	3	4	8	16
Control	34	42	49	56	67	77	118	219
Mucuna 25%	29	38	46	52	64	75	121	232
Canavalia 25%	25	33	37	41	46	51	76	139

Intake expressed in grams.

* Treatment – *Mucuna* is boiled in water for 30 minutes then dried and ground.

Source: Trejo, Belmar and Anderson unpublished data

Table 14 Performance of chicks fed diets containing different levels of *Mucuna* grain for 21 days

Inclusion level (%)	Average intake		Liveweight gain (LWG)		FCE*
	g %	s.e.	g %	s.e.	(g intake/g LWG)
0	2465 100	163	1372 100	173	1.80
14	2523 102	142	1224 89	144	2.06

28	2406	175	1030	154	2.33
	98		75		
42	1807	160	567	146	3.18
	73		41		

*FCE is food conversion efficiency. s.e. = standard error.
Source: Trejo, Belmar and Anderson unpublished data

Table 15 Short-term intake by chicks fed diets containing different levels of treated* and untreated *Mucuna* grain

Treatment	H O U R S							
(% inclusion level)	0.5	1	1.5	2	3	4	8	16
Control	12	14	17	20	29	37	57	117
28 % untreated	8	9	11	12	17	22	41	87
28 % treated	9	12	15	16	23	30	53	111
42 % untreated	6	7	9	9	13	18	32	64
42 % treated	6	10	12	13	19	24	43	88

Intake expressed in grams.
*Treatment: *Mucuna* grain lightly ground and soaked in water for 24 hours.
Source: Trejo, Belmar and Anderson unpublished data

Table 16 Performance of growing chickens fed different levels of *Mucuna* bean treated* and untreated for 14 days

Inclusion	Average intake		Liveweight gain		FCE**
level (%)	g	s.e.	g	s.e.	(g intake/g LWG)
	%		%		
0	1440	196	809	184	1.8
	100		100		
28 untreated	1473	172	601	100	2.5
	103		74		
28 treated	1476	135	623	103	2.4
	103		77		
42 untreated	1100	217	362	122	3.0
	77		45		

| 42 | 1245 | 182 | 442 | 111 | 2.81 |
| treated | 87 | | 55 | | |

*Treatment: Mucuna grain lightly ground and soaked in water for 24 hours.

**FCE is food conversion efficiency. s.e. = standard error

Source: Trejo, Belmar and Anderson unpublished data

The data in the tables above show that at 25 per cent inclusion in the diet Mucuna grain intake is comparable to the control diet, while the intake of the Canavalia diet was severely restricted (Table 13). This concurs with the findings of Belmar and Morris (1994). Voluntary intake of diets containing Mucuna is comparable to control diets at 14 and 28 per cent inclusion levels. At 42 per cent Mucuna inclusion intake is reduced to 73 per cent. Despite similar intake levels, liveweight gain is affected by increasing Mucuna inclusion (Table 14). Lightly grinding and soaking the Mucuna grain in water for 24 hours improves voluntary intake. From the short-term intake studies (16 hours) Table 15 shows that intake of diets that contain 28 per cent of treated Mucuna is comparable to the control diet, and intake of 42 per cent treated Mucuna is comparable to 28 per cent untreated Mucuna diets. By comparing Tables 13 and 14 we can see that there appears to be a threshold effect on intake of untreated Mucuna diets between 25 and 28 per cent inclusion. In longer-term intake studies (14 days) intake was affected above 28 per cent inclusion. The grinding and soaking treatment marginally improved liveweight gain compared with performance on untreated diets.

Table 17 shows the benefit of including Mucuna grain in the diet of female breeding pigs in terms of maintenance of body condition score, and litter performance at birth and at weaning. This data comes from the monitoring of homegarden pig keeping by campesino families.

Table 17 Field observations of small-scale pig husbandry systems, Calakmul, south-eastern Mexico: performance of gilts and their first litters

Parameter	Gilt's diet (with /without Mucuna)	Means	s.e.	n. (gilts or litters)	Prob. of difference between means
Gilt BCS of the gilt at 1st farrowing	with	3.06	0.14	35	0.37
	without	2.78	0.15	9	

Litter weights at birth (kg)	with	9.68	0.50	13	0.56
	without	9.19	0.66	10	
Number born alive	with	9.08	0.34	12	0.66
	without	9.40	0.69	10	
Gilt BCS at weaning	with	2.36	0.20	11	0.11
	without	1.90	0.18	10	
Weaning weight of piglets (kg)	with	5.10	0.18	10	0.42
	without	4.88	0.18	9	
Litter weight at weaning (kg)	with	45.17	2.16	10	0.33
	without	41.00	3.77	9	

BCS = body condition score (data from four communities, all the gilts were of imported breeds, they were offered diverse diets that contained combinations of maize, *Mucuna* grain and forage, sweet potato, chaya, and different forages).

Source: Anderson et al. (1997)

Conclusions

There is demand from farmers for cover crops to fulfil human food and livestock feed and forage provision purposes. Hence, there is a consequent need for researchers and development workers to be aware of this and to accommodate the food/feed/forage functions in cover crop interventions, particularly in food-insecure situations where cover crops can be used to alleviate the family/livestock competition for staple foods.

Until now there has been too little research and development emphasis on combining the provision of human food function of crops that can also be used for providing cover. Crops such as sweet potato have good food potential with cover characteristics.

The integration of cover crops and livestock in the same agricultural system can potentially provide significant economic and environmental benefits. However, there are various trade-offs to be considered. Many cover crops provide good quality forage but this function has to be balanced against the requirement for forage cover for other purposes. Unfortunately there are few cover crops that provide grain suitable for unprocessed feeding to monogastric livestock. The feed value of the grain is high once processed, but the costs (use of resources) of processing may be too high for poor rural families. There is evidence of some promising advances in terms of low-cost treatment of *Mucuna*. Other options for the improvement of cover crop grain feed value are breeding for reduced secondary metabolite content, or bioengineering to remove, or render ineffective, the anti-nutritional factors.

Cover crops also offer potential for the recuperation of degraded pastures. Certain cover crops can be used to smother invasive weed species, reduce soil compaction, improve soil infiltration, prepare the surface for pasture replanting, and produce (short-term) improvements in soil fertility.

Cover crops as a component of sustainable land husbandry

This chapter looks at the contribution of cover crops to sustainable resource use, with particular reference to aspects of the conservation of soil, water and species diversity. The forest/agriculture interface is used as an example, drawing on experiences presented at the Merida Workshop.

Farmer, national and donor priorities

For farmers, and particularly those with few financial resources, natural resource conservation is rarely an end in itself. Thus, any technologies designed to that end must also contribute directly – and preferably in the short term – to productivity, income and/or the reduction of risk. The exception might be in severely degraded situations where there is no alternative strategy to land rehabilitation (and there exists a support mechanism to allow that to happen while safeguarding minimal food security – e.g. food-for-work in soil and water conservation initiatives). Table 18 suggests a typical scoring of the importance of land-use options by major stakeholders.

Table 18 Land-use priorities for primary and secondary stakeholders

Stakeholder	Land-use purpose			
	Resource conservation	Soil and water conservation	Income generation	Subsistence production
Farmer	Low priority	Low priority	High priority	High priority
Government organization	Medium priority	High priority	Medium priority	Medium priority
Donor	High priority	High priority	Low priority	Medium priority

For donors, and to a lesser extent for governmental organizations, the various elements that make up sustainable natural resource conservation (soil and water conservation, maintenance of species and habitat diversity, environmental health, etc.) constitute an important component of aid policy and international commitments.

It is therefore rational for donors and NARES to promote the role of cover crops in soil and water conservation, watershed management and protected area/buffer zone management. It is equally rational for farmers to reject cover crops for these purposes unless they perceive short-term production, income or risk-amelioration benefits.

Soil and water conservation

Rapid depletion of soil fertility and non-sustainable land use are both a cause and a consequence of widespread poverty (ISCO, 1996). Cover crops contribute in three main ways to the conservation of soils:

- by contributing to the provision and cycling of soil nutrients
- by improving physical soil structure properties
- by reducing soil erosion.

The different factors influencing soil quality are interrelated, as presented in Figure 5. Cover crops can influence all of these factors, although the degree to which that influence will contribute to the functions required of cover crops by farmers and others will be location-specific, and dependent on the extent to which that factor is initially limiting productivity, profitability, versatility or sustainability.

Figure 5 Factors determining soil quality

Physical factors	Chemical factors
soil structure	soil pH
soil texture	
soil density	macro-nutrients
temperature	
erodibility	micro-nutrients
soil depth	

Water balance	Biological factors
drainage/leaching	humus
water retention	
water-holding capacity	soil fauna
groundwater level	
surface/subsurface	soil flora
movement	

Soil degradation processes represent an increasing threat for agricultural production and food security. The pressure on land resources increases with increasing population pressure. The term 'soil degradation' is used to describe a series of phenomena. In highly populated areas of Indonesia, where soil erosion is one of the most important factors leading to soil degradation, there are 8 million hectares of severely eroded land (Suwardjo et al., 1991). In Bolivia's Santa Cruz Department, the main soil degradation processes under mechanization are compaction, slumping, wind and water erosion, low levels of soil organic matter and nitrogen and weed infestation. Seventy-seven per cent of the predominant young and poorly structured soils that have been under mechanization for ten or more years are moderately to severely degraded (Barber and Navarro, 1994). In non-mechanized slash-and-burn agriculture in the same country, the principal soil-related problems are a rapid decline in fertility and organic matter after burning, low levels of soil nitrogen and phosphorus, waterlogging and weed infestation.

Depending upon the location-specific situation, cover crops take different functions in the process of soil conservation, contributing to reducing soil erosion (Thurston, 1997), improving soil organic matter, plant mineral nutrition and/or weed control. We do not attempt, in the following account, to prioritize the functions that cover crops have regarding soil conservation.

Soil fertility management

The maintenance of crop production requires the addition, recycling and retention of nutrients, as they are removed in the form of crop products during harvest and by the process of runoff, leaching and erosion. Different sources of nutrients can be identified: the mobilization of existing nutrients in the soil and parent rocks; the fixing of nitrogen from the atmosphere; or the supply of organic or inorganic fertilizer (Pretty, 1995). Different agricultural systems are based on different strategies of soil fertility management. The main categories are (Veldhuizen et al., 1997):

- shifting cultivation and fallow systems: soil fertility management is based upon the regeneration of natural vegetation (slash/mulch or slash/burn systems)
- interactive pastoral and fallow systems: nutrients are provided by animals and fallow vegetation
- integrated and organic agricultural systems: recycling of nutrients through composts, animal manure, crop rotations, etc.
- industrial agriculture: nutrients are incorporated in the form of inorganic fertilizer.

Inorganic nitrogenous and phosphatic fertilizer use has been promoted at all scales of agricultural systems, and has been adopted by farmers confronted with decreasing soil fertility. However, in situations where farmers are resource-poor, external inputs are often not affordable and farmers have to find other strategies to deal with the limitation presented by low or declining soil fertility.

The use of cover crops, especially nitrogen-fixing legumes, presents an option for the improvement or maintenance of soil fertility based on internally produced inputs. Legumes grown together with, or before, a cereal crop can reduce or eliminate the need for inorganic nitrogen sources. However, in many instances, the contribution of legumes to soil nitrogen status is disappointing. Rhizobial fixation can be reduced by poor availability of micro-nutrients and/or phosphorus, by the lack of the appropriate rhizobial strain (particularly for introduced species of legumes), and by physiological stresses on the fixation process such as waterlogging, drought or high soil temperatures. The contribution of rhizobial nitrogen to the growth of the legume and to associated or subsequent crops also depends on factors such as the amount of biomass off-take, and nitrogen losses due to leaching, nitrification and volatilization.

There is no equivalent to rhizobial nitrogen fixation that increases the total phosphate pool in the soil. However, cover crops can contribute to the increase of plant-available phosphate. Pretty (1995) states that recent research in semi-arid areas of India has shown that some legumes, such as chickpea and pigeon pea, have a mechanism that allows them to access phosphate in phosphate-poor soils. They release acids from their roots, which react with soil-bound phosphate to release phosphate for plant uptake.

Recommendations made by the USDA dating back to the beginning of the 20th century said that *Mucuna* was one of the most promising species for soil fertility management:

> The velvet bean (*Mucuna* spp.) is one of the best soil-improving crops, both for soils which are naturally infertile and for those which have become somewhat exhausted by long cultivation. The ability of this plant to make profitable growth on land so poor that most legumes will not thrive places it amongst the most important crops for the South [of the USA]. In addition to adding at a minimum cost large quantities of vegetable matter to the soil, thus making it more retentive of moisture, the nodules on the roots collect a large amount of nitrogen from the atmosphere. This nitrogen will be left in the soil when the crop is turned under and the plants decay. (Tracy and Coe, 1918)

Cover crops can contribute significant quantities of nitrogen through rhizobial fixation. However, this is often limited due to adverse conditions, e.g. waterlogging, poor choice of cover crop species, lack of suitable native rhizobia or the offtake of nitrogen as seeds or biomass. In some cases farmers might decide to incorporate leguminous residues, thereby enhancing their contribution to nutrient cycling and soil structure improvement (i.e. using the legumes as green manures), rather than leaving them as a living or dead protective surface cover (i.e. choosing to maximize the legumes' benefits as a cover crop).

Experiments carried out in Cuyuta, Guatemala, have shown that the nitrogen-fertilizer substitution value of *Mucuna* spp. and *Canavalia ensiformis* managed under zero-tillage (residues not incorporated) is around 60 kg/ha, whereas the value rises to 158 kg N/ha for *Canavalia* and 127 kg N/ha for *Mucuna* where the residues are fully incorporated (Moscoso and Raun in Barreto, 1991).

Farmers in Central America have experienced the benefits of relaid *Mucuna* on their maize yields, and the innovation has enjoyed widespread adoption (Avila and Lopez, 1990; Buckles in Barreto, 1991). NGOs have taken up the concept of *Mucuna*/maize cultivation stimulated by the successes reported from the region. A deficit of this promising and successful work in Central America is that the agronomic basis for these recommendations is rather anecdotal and not based on agronomic measurements (Triomphe, 1996). Very few of these experiences have been evaluated in quantitative terms, and most of the information available from the practice concentrates on farmer and extension worker testimonies.

The ways in which farmers and scientists evaluate soil fertility are different. Farmers' evaluation criteria are on appearance, manageability and productivity, while scientists base their evaluation on quantitative measurements of physical, chemical or biological properties. Each is valid, and the two are complementary, but they are often carried out independently from each other. Equally, the ways in which farmers and researchers evaluate the contribution of cover crops are also different. Each has much to offer the other in arriving at a more complete understanding of processes and positive and negative impacts.

A study carried out in Yucatan, Mexico, revealed different criteria used by *campesinos* to track the changes occurring with the integration of cover crops (*Mucuna*) and maize. Soil properties such as colour, texture, humidity and the potential to sustain demanding crops like chillies or tomatoes were identified as important by the participants. Changing soil colour was related to the existing soil types, which is the red-coloured *Kankab* and the black-coloured *Box luum*. The participants observed a change from the red colour towards a darker colour. Changes in the texture were experienced through: a) the ease of sowing and weeding, and b) by looking for organic matter to be found in the soils where green manure has been used. Increased soil humidity was related to the appearance of plants during water stress and was also detected by the soil temperature. Cold, or fresh, soils were related to more humidity, whereas warm soils were considered to be dry. As the *campesinos* traditionally use a patchwork of different crops according to different soil conditions, so the relation between soil quality and the capacity to grow crops like chillies and tomatoes, known to be demanding crops, was applied to soil evaluation (Gündel, 1998).

Soil structure improvement

Cover crops can provide large amounts of organic matter, which is a key aspect of soil structure management. Triomphe (1996) reports from northern Honduras that maize/*Mucuna* intercropping adds 30–50 per cent more soil organic matter to the 0–5 cm soil horizon than a comparable maize monoculture. The functions of organic matter in the soil are manifold; soil structure improvement is just one of them. Figure 6 presents the complexity of functions of organic matter production in land husbandry.

Figure 6 Functions of organic matter in land husbandry

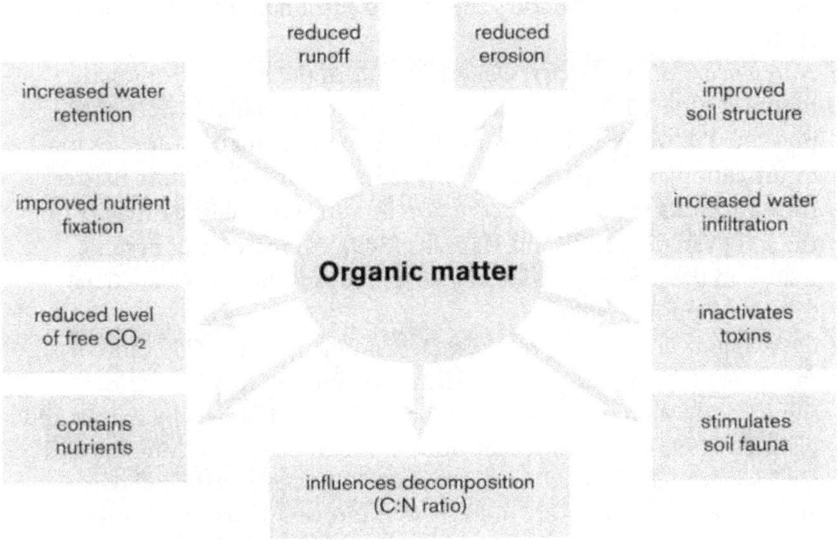

The following functions are related to the impact of organic matter on soil structure:

- *Water infiltration*: A consequence of reduced rates of water infiltration is that runoff losses increase. In hillside areas, water runoff leads to erosion and contamination of surface water sources. In low-lying areas poor water infiltration contributes to waterlogging. Organic matter improves the soil structure, and consequently the soil porosity and the capacity for water infiltration. Surface cover, as with a cover crop, improves the infiltration of water by reducing the speed of runoff. A strong root system creates channels for the water to enter the soil. Thus, by careful selection of cover crop species and planting date, the soil surface can be protected from damaging early- and late-season rains when water infiltration is crucial. A steady release of clean groundwater is essential to local health, and to that of distant populations. Runoff water typically has a high sediment and nutrient load, both of which are a problem to downstream reservoirs, whereas drainage groundwater is of better quality.

- *Soil water retention*: The soil's capacity to retain water is crucial to efficient utilization of available water. Soils with a low organic matter content have less capacity to retain water than those rich in organic matter. For rainfed agriculture, this characteristic is very important, as the crop development depends upon its access to soil water. Farmers in Yucatan have a local soil classification system, which includes among many other criteria the soil's capacity to sustain plants during dry spells.

- *Nutrient-holding capacity*: Generally soils in the humid tropics are highly weathered and leached, with most of the available nutrients presented in the organic topsoil. Because organic matter supplies most of the cation exchange capacity, any reduction in soil organic matter (SOM) results in a marked reduction in nutrient-holding capacity of the soil (Van der Heide and Hairiah, 1989). An increase of organic matter in the topsoil provided by cover crops is therefore beneficial for the cropping system.

- *Decomposition of organic residues*: Depending upon the C:N ratio of the cover crop species, the long-term contribution to SOM varies significantly. A study of 14 cover crops in lowland Bolivia by Barber and Navarro (1994) concluded that while cover crops such as Mucuna and *Dolichos lab-lab* supplied large amounts of nitrogen in their above-ground residues, they were unlikely to improve soil organic matter content or subsoil structure, because of the low C:N ratios of the above-ground residues and the scarcity, small diameter, and low C:N ratios of their roots. In contrast, the grasses *Brachiaria brizantha* and *Panicum maximum* (varieties Tobiata and Centenario) might be expected to improve SOM and subsoil structure significantly due to the medium C:N ratios of above-ground residues, and moderate-to-high density of roots of high C:N ratios at 15–30 cm soil depth. Pigeon pea (*Cajanus cajan*) supplied large amounts of nitrogen in the above-ground residues, and might be expected to improve subsoil structure due to the large diameter of its roots of medium C:N ratio.

- *The pattern of release of nutrients* from cover crop residues, and its synchrony with the demands of crops, is not well documented. However, it is known that some cover crop residues decompose quickly after cutting or leaf fall, whereas others decompose much more slowly due to their lower leaf to stem ratios, higher C:N ratio, or the presence of chemicals such as tannins which inhibit bacterial breakdown. These characteristics are important criteria to bear in mind when selecting cover crops for particular functions such as fertilizer substitution, weed suppression, organic matter build-up or erosion control. Further classification of cover crops for these characteristics would be useful to researchers and farmers alike.

The extent to which farmers are aware of these concepts, and use them in taking soil management decisions, is poorly understood. The promotion of appropriate soil management strategies depends upon the sharing of perceptions, information and concepts – including the principles behind the potential benefits of cover crops.

Erosion control

Soil erosion is a phenomenon whose severity is determined by several factors: the location (slope), the strength and duration of rainfall events, the management practices (including, crucially, soil cover) and the soil structure. Approximately 25 per cent of land in Latin America is on hillsides and plateaux that are susceptible to erosion. Such areas are usually inhabited by poor Indian populations whose livelihood is threatened (Alamgir and Arora, 1991). Soil erosion is a progressive process, which passes through various stages until it results in a complete loss of production capacity. This critical threshold point is reached when the soil layer has decreased to a point where plant roots are unable to develop. Studies in Indonesia, where land degradation is a serious problem in upland agriculture, have shown that the rehabilitation of this land requires at least one year under a cover crop (Suwardjo et al., 1991). The more serious the erosion, the longer the rehabilitation period. Rehabilitation attempts promoted by the Indonesian government using tree species have failed because farmers prefer to grow food crops, even on steep land. Another attempt to overcome the problem of degraded land was to study the potential of five different cover crops. *Setaria* spp.,

Crotalaria spp., *Pueraria* spp., *Centrosema* spp. and *Psophocarpus* spp. have been tested for two years on degraded soil in West Java. The results showed that *Centrosema* had the greatest potential to improve soil productivity. However, farmers did not adopt this cover crop species, because they do not obtain a direct benefit (food or income) from the crop. Taking this criterion into account, new trials were established using legumes which are considered to be edible. These legumes are cowpea, *Mucuna* (*Mucuna* is appreciated as an edible legume crop in Central Java), *Dolichos lab-lab* and mungbean. Results of this trial show that *Mucuna* had the most positive effect on soil physical properties due to its higher biomass production (9.2 t/ha) compared to cowpea with 5.9 t/ha; *Dolichos lab-lab* (4.5 t/ha); and mungbean with 2.3 t/ha (Suwardjo et al., 1991).

Experiences from Merendon reserve in Honduras

The following case study, presented by Fürst (1997) from Honduras during the Merida Workshop, describes an experience of soil and water conservation from several communities of the mountainous area of the Merendon reserve in northern Honduras.

Box 17 The Merendon experience of soil and water conservation

About the region: The Merendon mountains in northern Honduras can be divided into three zones based on their geographical and climatic characteristics. The lowland area is located between 300 and 700 m and receives 1300–1900 mm rainfall per year; the second area is located between 600 and 900 m with 2500 mm of annual rainfall; and the third area lies between 900 and 1400 m with 3000 mm rainfall. The small-scale production system of the Merendon *campesinos* includes the cultivation of coffee, basic grains, vegetables, flowers and, less frequently, livestock. Soil erosion and decreasing soil fertility are major constraints to agricultural production.

The FUNBANCAFE initiative: The Honduran NGO, FUNBANCAFE is working with 19 *campesino* communities in this region. The project provides technical assistance, training and education to more than 900 families within these zones, and the main objective is to promote technologies of sustainable agriculture which allow the protection, recuperation and conservation of natural resources, and the improvement of livelihood conditions and community organization. Parts of the area are declared as natural forest reserve. Apart from this, the Merendon area is the extraction zone for drink-

---

ing water for the nearby city of San Pedro Sula. The interest of the local government is to ensure an adequate water supply in terms of quality and quantity.

As soil erosion and decreasing soil fertility are major constraints to agricultural production, the project seeks to promote sustainable agriculture technologies, including agroforestry, cover crops and contour planting. Cover crops have been established in crop rotations, intercropped with basic grains and used as ground cover in the coffee plantations. Genera used include *Mucuna*, *Canavalia* and *Dolichos*.

The methodology used to promote these technologies is through local volunteer educators, who are collaborating with the project extension staff. Field visits, technical manuals and the distribution of seed material are some of the activities used to diffuse the innovations. The project is guided by the following two affirmations: 'It is not the agriculture we wish to transform, but rather the people who are practising the agriculture,' and 'It is not the concept of soil erosion we have to focus on, but rather the concept of soil as a living organism and its fertility.'

The following observations have been made by the project staff regarding the use of cover crops in the project region:

Positive observations:	Limitations identified:
o increase in maize production	o cover crops show foliage fungus diseases in the high, humid zone
o reduced soil erosion	
o soil structure improvement	o seed production of cover crops is poor in the high, humid zone
o recuperation of degraded areas	
o decrease in weed infestation	o in heavily weed-infested areas it takes up to 2–4 years to improve cropping conditions
o reduction in herbicide use	
o increased rearing of livestock	
	o some *campesinos* fear that the cover crop mulch on steep slopes might lead to landslides
	o maize harvest is more difficult in plots with cover crops

The project also identified some limitations in terms of adoption of cover crop innovations. A serious limitation perceived by the project, as well as by the farmers of the area, is the land tenure system. It is difficult to motivate the *campesinos* to establish cover crops for soil conservation in plots that are only rented. They suspect that the owners will reclaim their land once the soil conditions have improved. Therefore *campesinos* with access to their own land are more likely to benefit from cover crop technologies than are landless *campesinos*. A further limitation is the restricted availability of cover crop seed material, especially in the zones where seed production is difficult due to high humidity. In the future the project seeks to develop alternative cover crop technologies using local *Vigna* and *Phaseolus* varieties.

Issues arising from the case study

The following important points can be summarized from the case study and from the Merida Workshop discussion group:

Time-scale of cover crop interventions for soil improvement
Cover crops are not an instant solution to problems like soil erosion, even though soil loss and surface runoff can be considerably reduced from the first season of use. It takes time to adapt the technologies to local conditions and to build up seed stocks to be able to continue the process. Degraded land takes several years to recover its fertility and structure. The effective management of aggressive weeds such as *Imperata* can take several years to achieve. This is often contradictory to farmers' direct needs and problems, which require short-term solutions. Working with farmers is essential, and it is necessary to demonstrate economic and productivity benefits before spontaneous and widespread adoption is assured. The different agendas of farmers, government and donors for promoting and using cover crops need to be recognized, so that realistic expectations are predicted and met.

Complementary practices
While cover crops contribute to soil erosion control through surface cover, the addition of SOM and increased infiltration on steep slopes are sometimes not sufficient on their own. In these cases, additional measures such as live barriers, terraces, contour lines (e.g. *Vetiver* or crop residue 'trash lines') are required, in conjunction with the control of grazing and burning.

The socio-economic environment
Cover crops provide a partial technical solution to soil degradation. Often cover crops need to be linked to other measures, such as controlled grazing, agroforestry or reduced burning. Without an enabling and supportive socio-economic environment, successful application is unlikely. Outsiders

can sometimes encourage community reflection about a worsening situation. Land tenure issues present a serious limitation to the adoption of long-term measures, where the local situation discourages individual investment in land-use improvements. In some cases, community action can be stimulated for improving communally-held property, depending on the degree of cohesiveness in the community, the shared appreciation of the need and likely benefits and the confidence they have in support from external organizations.

Towards a shared understanding of the situation
With cover crops, as with other technologies, dissemination methods either try to supply a recipe to be followed, or assist the development of an understanding of the principles behind technologies in order that they are applied and adapted appropriately. As cover crop functions regarding land husbandry tend to be less tangible, and longer term, than those relating directly to productivity (such as fertilizer use), it is important to share the different concepts of external organisations and farmers on the potential changes and impacts to be achieved through the use of cover crops. A mutual learning process is required to identify appropriate management strategies. Hidden agendas, like the improvement of water quality for urban settlements, or the reduction of siltation of distant reservoirs, will not contribute to the local uptake of new technologies (unless incentives are given that are transparently linked to these agendas).

'Above-ground' natural resource management

Degradation of natural resources occurs not only in terms of soils ('below-ground') but also in terms of natural vegetation (forests) and cultivated crops ('above-ground' resources). Worries about the loss of natural forest areas due to the constantly advancing agricultural frontier, the increase of chemical use within the ecosystem and the decrease in genetic crop diversity are issues of widespread political and societal concern. Cover crops are promoted in this context as a component of alternative approaches to agricultural practices that tend to deplete those resources. The situation at the forest/agriculture interface is taken as an example, for which different experiences were presented at the Merida Workshop.

Cover crop technologies at the forest/agriculture interface

Shifting cultivation, characterized by the slashing and burning of forest or bush-fallow to provide a temporary seed bed for annual crops, is a common agricultural practice in the humid tropics of Africa, Asia and

Latin America. Shifting cultivation includes a variety of different traditional practices that have evolved under local circumstances as well as those used by colonizing farmers. The latter are less sophisticated in their management of natural resources, often leading to the rapid degradation of forest and soil resources. These different practices rely on the fertility released by burning vegetation to sustain annual crops over one or two seasons, after which adverse soil or weed conditions require farmers to return the land to fallow. The fallow period is typically at least five times the cropping period for a restoration of fertility and the smothering of gramineous weeds.

Shifting cultivation systems have survived for centuries, but increased population pressure, changing aspirations of forest dwellers and the exploitation of tropical forests for logging, ranching, industrial agriculture, charcoal, settlement, national parks and reserves and other uses, has affected the once sustainable practice of shifting cultivation. 'Millions of poor farmers have turned to the ancient systems out of necessity, and its inherent sustainability is succumbing to these pressures' (Peters and Neuenschwander in Thurston, 1997).

The use of cover crops in these environments can lead to a crucial change in the conventional cropping practices by substituting the bush-fallow by a permanent mulch management practice. Experiences presented at the Merida Workshop have shown that cover crops can lead to a transition process from annual rotation systems towards permanent land-use practices. Perennial crops were incorporated into the cropping pattern, leading to an association of annual and perennial crops. The emerging crop arrangements can be considered as agroforestry systems, seeking to diversify spatial and temporal use and function of an area in order to optimize resource use, diversify income and product sources, and minimize the risk of soil erosion (Lal, 1990).

The following two boxes present examples from Nicaragua and Mexico, where cover crops have been promoted at the forest/agriculture interface. The first case (from Nicaragua) will be presented in more detail in Chapter 5, where aspects of the diffusion approaches used in Bosawas are discussed.

Box 18 The Bosawas experience (Part 1)

Background: The Atlantic region of Nicaragua contains the largest tropical humid forest reserve in Central America; 7500 km² have recently been declared as a protected area, known as the Bosawas Reserve. As elsewhere, this protected area is threatened by the advancing 'agricultural frontier', among other factors. The migration of *campesinos* searching for new and fertile land represents a phenomenon that is difficult to control and to regulate, because it is a survival strategy for many rural families.

The National Union of Agricultural Producers (UNAG) Initiative: Over the past three years UNAG has been working with *campesino* communities in the area, attempting to develop alternative management practices for natural resource use. The emphasis of their work has been on *campesinos'* participation in order to develop an experience that is led by the *campesinos* themselves, based on motivation, experimentation and promotion.

The process of change: This participatory process has led to a more rational land-use husbandry, where dual-purpose crops like *Cajanus cajan*, *Vigna* spp. and *Phaseolus vulgaris* have taken important roles in addition to *Mucuna* in improving staple crop production. The shifting cultivation practice has been gradually replaced by a permanent land-use system. This change has allowed the integration of perennial crops into fields where beforehand only annual staple food crops like maize, rice and beans have been grown. Crops like papaya, bananas, pineapple, and also tree species for multiple uses, have been established in combination with annual crops, leading to the diversification of products and the development of innovative agroforestry practices.

The following testimony of Jesús García from Rosa Grande illustrates the perspective of the *campesinos* participating in the UNAG project: 'We don't want to continue cutting down the forest vegetation now that we have seen the advantages of sowing *Mucuna* in our plots. This plant is very good to recuperate and protect the soils and to increase maize yields. The yields in plots with Mucuna are equal to the ones we have in newly cleared spaces.'

The possibilities presented by the use of cover and dual-purpose crops are the following:

- reducing deforestation by substituting the bush-fallow phase by cover crops
- more production from less area
- selection of areas for natural forest regrowth
- establishment of perennial and semi-perennial crops like pineapple, banana, coffee, cacao, etc.
- planning of permanent land-use systems.

Source: Rivas and Zamora, 1996.

Case study: Yucatan, South-east Mexico

A similar experience to the Nicaragua case has been reported by an action-research group from the University of Yucatan, Mexico, working with *campesinos* and several local NGOs on the participatory development of innovative technologies for sustainable resource use. The circumstances have already been introduced in Chapter 2, but now special emphasis is given to the potential of cover crops as a component of alternatives to slash-and-burn agriculture. The main focus of this case is on the integration of the introduced technology within the overall agricultural system, and the changes provoked by this innovation.

Box 19 The *milpa* system in Yucatan

Background: The Mayan *campesino* families of the Yucatan Peninsula in south-east Mexico have traditionally depended upon a slash-and-burn system of staple food production, locally known as *milpa*. The *milpa* depends on the forest as the main (fertility) input into its production system. For more than 3000 years the *milpa* formed part of an integral forest management system, which was characterized by rotation between forest clearance, agricultural use and forest regrowth. Nowadays productivity is constrained by the scarcity of land left fallow for sufficient time to recuperate fertility.

Alternatives to slash-and-burn: As a reaction to this situation, several local NGOs working in agroecological development made an effort to introduce into rural communities a legume-based cover crop technology as an alternative to the traditional *milpa*. The system is known in Yucatan as '*labranza minima*', which means 'minimum tillage'. An important potential of the innovative system is its contribution to a gradual shift from rotational land use towards permanent land-use management. The tendency identified in the adaptation process of the innovative system, shows that the participating *campesinos* have started to develop a long-term perspective on land use which is leading to the integration of perennial crops close to the minimum tillage plots.

The following map, developed by a group of *campesinos*, reflects their future vision for land use.

Don Dios from Sahcaba explains his vision of land use:

> The map shows our ideas of how we want to use our land in the future. Since *labranza minima* offers the possibility to have a permanent plot in one area of the *ejido*, we are planning to establish crops which grow over several years as a border to the *labranza* plot. For instance, we can use the periphery areas of the *labranza* for fruit trees like papaya or banana and fodder trees like *Huaxim* (*Leucaena* sp.)...

Figure 7 Future vision of the *milpa* agricultural system

Source: Campesino group from Sahcaba, 1996 (Campesino Workshop)

1 = *labranza* or minimum tillage; 2 = permanent *milpa* with *Mucuna* (managed without burning); 3 = *Mucuna* monoculture for animal fodder production and land improvement; 4 = perennial crops established in the periphery of the permanent plots; 5 = pigs kept in confined areas within the homegarden fed on *Mucuna*, maize, etc.; 6 = homegarden; 7 = fodder trees established in the homegarden (*Leucaena*, etc.); 8 = *campesino* family and accommodation; 9 = chicken hutch.

Results reached so far: The positive aspects in terms of sustainable resource management (forest vegetation) are:

- The innovation has the potential to reduce the amount of forest cleared for the *milpa*, as it contributes to household maize production and absorbs family labour resources usually used for forest clearance.
- The areas dedicated to the innovative technology have not been burned since the year of establishment.
- The experimentation process has facilitated an awareness creation among the participants of the importance of maintaining trees in their system.

On the other hand, there are factors that present a strong challenge to forest preservation and renewal:

- The present communal land tenure system does not give incentives for the establishment of forest areas. Land use is characterized by common access and use; thus other *campesinos* can cut the forest that has been left to grow by *campesinos* using the *labranza* system.

- Unless the *campesinos* perceive an economic value arising from protected forest areas it is unlikely that the conservation of forest will become a priority.
- Government programmes are often contradictory to the efforts made by local NGOs and *campesino* groups. They provide subsidies and credit for agricultural production systems which do not take into account the environmental aspects – nor do they promote sustainable management practices.

Implications for the future: It is likely that some community members will be interested in a permanent land-use system, whereas others will wish to continue with the traditional rotation. Hence, it would be necessary for the community members to agree on a new format for land-use planning that considers the needs of both.

Community-based land-use planning and the increased number of people involved in permanent (non-slash-and-burn) agriculture, offer the opportunity to define a contiguous area within the permanent section of the communal land, which can be dedicated and used as forest reserve. The ecological value of such a forest area would be greater than fragmented individual plots (Gündel, 1998).

Conclusions from the case studies

The important points that can be concluded from the two case studies are the following:

Towards permanent land management
The introduction of an innovation such as cover crops can be the key to profound long-term consequences, such as the gradual shift from rotational land use to permanent land management.

Complementarity to existing practices
Innovations, such as minimum tillage with cover crops or the sole incorporation of cover crops into staple food production, can be complementary to traditional or existing systems and provide an alternative for those situations where its advantages outweigh those of the traditional system.

Involvement of different stakeholders
Communal land ownership can limit the potential of cover crops as a component of permanent land management, unless all relevant stakeholders are involved in the process. Successful adoption at community level requires negotiation across the whole community.

Maintaining crop diversity

Farmers practising LEIA in marginal areas have been conserving significant amounts of plant genetic diversity, at both the species and intraspecies levels (RAFI, 1998). They depend on varietal mixtures, multiple crops, intercropping, homegardens and polycultures, as well as on genetically diverse varieties of individual crops. This fits with the farmers' conditions and their need to cope with high variability in their edaphic, climatic and biological environments, and their limited access to, or inability to purchase, inputs (Iwanaga, 1995). However, these agricultural systems are subject to external pressure and changes, which often result in the loss of crop genetic diversity (Cooper et al., 1992).

Different factors contribute to this process: market forces, creating new preferences, migration and demographic pressure, pressure from extension agencies, formal education and cultural change, technological change and economic development, land degradation, soil erosion and long-term environmental changes (Cooper et al., 1992). Reduced crop diversity can lead to soil degradation that affects landscape-level processes. For example, the use of monocultures can reduce soil moisture-holding capacity, which allows greater leaching and surface runoff of soil particulates, nutrients and pesticides. Less crop diversity may also result in greater crop losses to pests (insects, weeds and diseases) in which the dispersal of pests is enhanced, the rate of adaptation to crops by pests is accelerated, and the associated resistance of crops to pests reduced (Barrett et al. in Francis, 1994).

Table 19 shows the drastic reduction of crop varietal diversity of beans and peas, two species with important potential as cover crops.

Table 19 Reduction of crop biodiversity in beans and peas on a global level

Common name	Latin name	Number of varieties		Loss %
		1903	1983	
Beans	Phaseolus vulgaris	578	32	94.5
Peas	Pisum sativum	408	25	93.9

Source: adapted from McNeely, in Engels (1995)

During the Green Revolution, high-yielding varieties have been promoted, which have expelled many of the traditional varieties. Kothari (1997) mentions the Green Revolution strategy as an important factor that has caused a dramatic change towards monoculture in Indian agriculture in the last few decades. Other interventions, like the establishment of irrigation schemes, have homogenized previously diverse micro-habitats, due to a sub-stitution of intercropping by monocropping.

The promotion of high external input agriculture is not the only cause of the loss of crop diversity. Concerns regarding the enthusiasm for *Mucuna* as a suitable cover crop have been expressed by various sources. The following contribution made by the MISEREOR Foundation during the Merida Work-shop highlights the possible negative effects of over-emphasizing *Mucuna*.

Box 20 Traditional cover crop legume species in South Honduras

In the late 1960s or early 1970s *Mucuna* was introduced as a cover crop in Guatemala and Honduras. Due to its abundant organic matter production and its vigorous growth, *Mucuna* spread among farmers in the northern coastal region. Later on, this technology was taken up and promoted by many govern-mental and non-governmental organizations throughout the country. This process has led to a situation in some communities in the south of Honduras where *Mucuna* has become a monoculture, presenting the same risks as any other crop cultivated in monoculture, including susceptibility to pests and dis-eases, and risk of total loss under drastic climatic conditions. This massive introduction of *Mucuna* happened in an area which had a large variety of local legume species. They are generally intercropped in the local maize production system (*milpa*) to provide staple food and income generation. In combination with maize, sorghum, sweet potato and other crops, these legumes have a potential role as cover crops, which is not taken into account by researchers and development agents. The main criticism regarding the introduction of *Mucuna* is that it has become a technology package, which is imported into a local situation without taking into account the potential of locally available crop species.

Advantages of the local legume species include:

- local legume species provide a soil cover without being as abundant and aggressive as *Mucuna* tends to be
- they are well known within the communities
- some can be used for human food or as a market crop
- some of them can be consumed as green pods, providing food at critical periods of the year.

Source: Jacqueline Chenier, MISEREOR (1997).

In addition to the favourable points listed by Chenier we should like to add the aspect of the adaptation of local species to the environmental conditions. Many introduced cover crops suffer from poor development because of adaptation problems. Recent experiences from the Kathmandu Valley in Nepal, aiming at the development of Sloping Agricultural Land Technologies (SALT), show that introduced Faba beans (*Vicia faba*) were in poor condition and not nodulating in most of the cases analysed. The same observation was noted for introduced white clover and peas; meanwhile, native clover weeds and a local kitchen-garden faba bean were growing quite well (Keatinge et al., 1997).

Notwithstanding these advantages, it must be acknowledged that local legumes are not fulfilling all requirements. The reasons for this, and the potential contribution of introduced species, should be considered carefully and without bias.

Research by CIAT/NRI in Bolivia has started to evaluate local landraces of cowpea (*Vigna unguiculata*) as cover crops for the winter season when temperatures are cool, and some moisture stress is expected. Some landraces have an indeterminate habit, covering the soil for some five months, and producing small amounts of seed for food over an extended period of the hungry (dry) season.

Potential of local cover crop species

The choice of cover crop species should not automatically be from the small number of well-known species, but should consider local species that may have greater adaptability to physical conditions and local needs. The farmer's role is to explore and to experiment with those local species, providing an important source of information on their potential functions as cover crops.

Widening the perceptions

There is potential to select within local landraces for crops that can fulfil cover crop functions, even though they had not previously been perceived in that role (e.g. local indeterminate landraces of cowpeas in Bolivia that can function as dual-purpose food/cover crops). This remark is valid for the different stakeholders involved in the promotion of cover crop technologies. Farmers might not perceive the cover crop function of a food legume traditionally grown in his plot, and likewise extensionists and researchers may not consider it. A learning process from the different stakeholders is required which allows an analysis of the existing resources to fulfil different functions.

Reduced use of agrochemicals

Cover crops potentially reduce the use of certain agrochemicals, and the possible contamination of the environment through misuse (e.g. accidental runoff, spillage or drift). Those leguminous cover crops that are net nitrogen producers can substitute in part for nitrogenous inorganic fertilizers, while those that give rapid and complete soil cover can reduce the need for herbicide use. These environmental benefits (of interest to donors) also benefit farmers directly, as they represent a reduction in cash outlay. At the governmental level there is the benefit that foreign exchange, for the importation of inputs, is substituted.

Conclusions

- Cover crops do not present a short-term solution to the degradation of natural resources – soil, water, etc.
- Soil and water conservation requires strategies that combine different actions which might include the incorporation of cover crops.
- Land tenure issues and other socio-economic aspects have to be taken into consideration before the successful promotion of cover crops can be achieved.
- Land husbandry involves many different stakeholders. To achieve a more sustainable land husbandry these stakeholders have to be participants in the development and evaluation of innovations.
- The identification and promotion of local cover crop species can contribute to an increased crop diversity in agricultural systems.

5

Farmer experimentation and diffusion strategies for cover crop innovations

Farmer experimentation and diffusion strategies were regarded as crucial issues by the participants of the Merida Workshop. The different experiences presented during the workshop showed that the cover crop innovations currently being promoted had been subject to informal experimentation, adaptation and diffusion by farmers. Farmers' involvement was facilitated through the participatory approaches used by the different agencies involved. Innovations that offer a range of management options allow farmers to take an active role in the process of adapting them to their specific conditions and needs (Gündel, 1998).

The first section of this chapter presents two case studies (from Nicaragua and Honduras) where farmers took an active role in the adaptation and diffusion of cover crop technologies.

The campesino-to-campesino approach

The origin of the *campesino*-to-*campesino* approach lies in the 'Popular Education' movement which developed during the 1970s and 1980s in Latin America. Equitable communication and mutual learning processes, combined with 'reflection–action–reflection' methodologies for social transformation and local empowerment, effectively put control of the learning process in the hands of local people. In Central America thousands of promoters have become involved in developing literacy, health and community organization (Holt-Gimenéz, 1996).

In agricultural projects that have taken this approach, it is *campesinos* who assume the role of leaders in experimentation, innovation, training and diffusion, supported by external technicians in the role of facilitators. Locally selected *campesino* 'voluntary promoters' are chosen for their qualities of commitment, motivation and willingness to share. In some cases 'part-time' paid *campesino* promoters (or 'para-professional' extensionists) have been used (Farrington, 1995).

The horizontal communication network between *campesinos*, technicians and promoters is meant to facilitate the spread of innovations from *campesino* to *campesino* in order eventually to reach the whole community (Okali et al., 1994).

In the next section a case study is presented from Nicaragua, where farmers have taken an integral role within innovation development and diffusion processes of cover crop technologies. Part 1 of the Bosawas experience has already been presented in Chapter 4. This second part provides information on the '*Programa Campesino-a-Campesino* (PCaC)' developed and promoted by the National Union of Agricultural Producers (UNAG).

Box 21 The *Campesino*-to-*Campesino* movement in Bosawas, Nicaragua

Farmers' organization in Nicaragua: During the Sandinista decade (1970–80) UNAG was formed. UNAG provided, and continues to provide, an important structure for communication, training and commercialization for agricultural producers, including smallholders and *campesinos*. The organization has provided the infrastructure for the PCaC programme, an initiative which involves thousands of *campesinos* all over Nicaragua. PCaC is working in a post-revolution environment, facing the problems of a divided society. Taking this division into account, the programme seeks to work on a basis that emphasizes the 'things which unify the people and not those which divide them'.

The PCaC proposal: The initial PCaC proposal was to set up a programme of soil conservation directed toward small-scale farmers, including the training of voluntary extensionists selected from among the participating farmers, who would be able further to diffuse a process of agricultural transformation through training. Taking this as a basis for its work, PCaC has gradually developed a methodology which:

o   respects the needs of the individuals involved
o   seeks to identify and to overcome constraints
o   makes use of local resources maintaining an ecological perspective
o   transforms the former top-down relationship between extensionists and farmers.

The initiation of PCaC in Bosawas: Siuna, a municipality in the south-east of the Bosawas reserve is characterized by a rapidly growing population that has advanced into the Bosawas reserve. Rosa Grande, a community in which PCaC initiated its activities in the region, comprised only two families in 1955. Local farmers have described how this number increased drastically over the

---

years, and reached 251 families in 1996. Most inhabitants are migrants from the humid zones of the Atlantic coast where traditionally they cultivated crops in slash-and-burn agriculture.

Most of the families in Rosa Grande work their own land. Seventy-one per cent have between 20 and 60 manzanas (equivalent to 15–40 ha); only 8.4 per cent possess more than 70 manzanas; 12 per cent own less than 20 manzanas, and 8.4 per cent own no land. Those that have larger plots cultivate their land under slash-and-burn agriculture within the limits of their own land holding. The 12 per cent who own less than 20 manzanas need to cultivate on national land, or have to rent land from other farmers. It is this last group that has been prioritized for participation in the PCaC, because it is they who are involved in the advance of the agricultural frontier. The group of landless families either borrow a piece of land from neighbours, or they enter the reserve. This has become more and more difficult because of increasing controls.

With five to six thousand farmers affiliated to UNAG, Siuna is one of the most successful regions in organizational terms. The *campesino*-to-*campesino* programme was initiated there in 1993. An exchange visit took place between a group of farmers from Rosa Grande and some other communities from Siuna, and communities elsewhere in Nicaragua where PCaC had already promoted cover crop innovations successfully. The aim of the exchange visit was to provide the *campesinos* from Siuna with new information on alternative land-use technologies and to motivate an exchange of plants, seeds and ideas between the participants. The ease of communication between the farmers, due to their shared language, facilitated an understanding of the functions and purposes of the cover crop innovations. The *campesinos* from Siuna returned to their communities and put into practice what they had learned from their visit.

Based on farmers' initiatives to try out new cover crop technologies, a multiplication process within the community and between communities was initiated. Exchanges, workshops, visits to individual farmers and field days were among the instruments used to enhance the diffusion of the innovative technology.

Jesús García, a *campesino* from Siuna, reflected:

> With the *campesino*-to-*campesino* methodology we have multiplied our experiences within the municipality. Between 1993 and 1995 we received more than 300 *campesinos* to our fields where we use the fertilizer bean (*Mucuna* spp.) In 1994, 1995 and 1996 *campesinos* from various communities started to use the fertilizer bean and there have been many training workshops based on our experiences. As a result more than 3000

campesinos have participated in field visits, training workshops and community exchanges.

Farmers' experimentation

This process has motivated many *campesinos* to put into practice what they had seen first in neighbours' fields. A dynamic has been generated of farmers' experimentation and appropriate, locally adapted innovation development. Table 20 shows some of the different experiments planned and carried out by the *campesinos*, and the reasoning behind them.

The table gives the name of the *campesino*-experimenter, his community, descriptions of the experimental procedures, and what their objectives were. The majority of the farmers' experiments were to investigate the effects of combining different crops and legumes.

Table 20 Farmers' experiments with cover crops in Siuna region

Farmer	Community	Experiment	Objectives
Fabián Saavedra	Siuna	Sowing legume cowpea (*Vigna unguiculata*) in May and cut it back at the flowering stage to sow rice in July	Find out if it is possible to plant rice every year in the same plot using legumes. Test potential legumes other than *Mucuna*
Jesús García	Rosa Grande	Sowing maize in *Mucuna* mulch at different densities	Observe differences in yield
Pedro Mairena	Aló Central	Sowing sorghum into *Mucuna* mulch	Evaluate the development of sorghum in this zone
Emilio Arosteguí	Las Brisas	Sowing melon and cucumber in plots that have been cropped with *Mucuna* for two years	Observe yield responses
Gregorio Orozco	Hormiguero	Sowing maize into *Mucuna* mulch	Observe yield effects
Andrés Romero	Yaoya	Sowing rice into *Mucuna* and *Canavalia ensiformis* mulch	Observe the development of rice
Eufracio Calderón	Danlí	Sowing *Canavalia* and *Mucuna* in May, cutting back in October and planting maize into the mulch	Find out which mulch results in better crop response

Agustín Mendoza	Santa Fe	Sowing *Mucuna* into the maize stubble, and cut it back in October to sow the second maize crop	Observe yields and improvement of soil fertility
Julio Jarquín	Las Brisas	Sowing of mungo (*Vigna mungo* or *Vigna radiata*), cut it back when flowering and sow maize	Compare yields with *Mucuna*
Carmelo Ubeda	Danlí Arriba	Sowing oats and after harvesting establishing red beans (*Phaseolus vulgaris*) into the oat stubble	Find out if oat residues improve bean yields

The outcome of the experimentation process was a range of individual experiences and new management alternatives, which were shared by the *campesinos* and which motivated other farmers to find their own solutions to their problems. The experimentation process and the farmer visits helped to maintain the links and communication between the farmers involved, and contributed to the identification of local promoters.

Table 21 summarizes the activities carried out in Siuna in 1996.

Table 21 Diffusion activities carried out in Siuna in 1996

Activities	Number	Total participants
Training Workshops on the use of *Mucuna*	16	320
Exchange of experiences within Siuna region	30	600
Promotion sessions using videos on *Mucuna* use	35	1400
Exchanges outside Siuna region	10	69
Total	91	2389

It is important to mention here that during the two-and-a-half years of activities, no full-time technicians were employed.

Adoption process of Mucuna
Figure 8 represents the adoption process of *Mucuna* in Rosa Grande, a community in Siuna region.

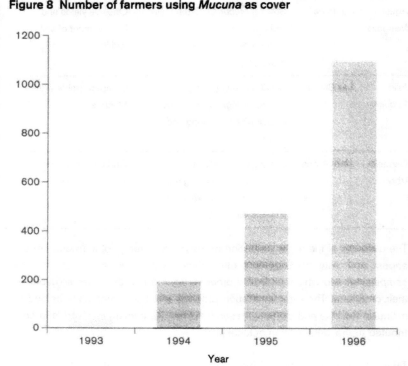

Figure 8 Number of farmers using *Mucuna* as cover

As the programme developed, not only the number of farmers using cover crops increased, but also the area dedicated to cover crops by individual farmers expanded.

In 1998 there were more than 2000 farmers in Siuna who were using or experimenting with *Mucuna*. Seventy communities within the Siuna region were involved. As a result of the *campesino* exchange visits the innovation also spread to the north-east parts of the Bosawas reserve, where other farmers from 15 communities started to experiment with the innovation.

Source: UNAG case study

Discussion points:

- Inter-community and intra-community diffusion: is the increase in the adoption rate caused by the diffusion between communities, or is there also significant diffusion within communities?
- Abandonment: do farmers discontinue the use of *Mucuna* once they have tested the innovation? And if so, why?

- Importance of flexibility both in the technology promoted and the way the promotion takes place.
- Embedding the promotion within the farmers' empirical learning process.
- The importance of the shared innovation process is perhaps greater than the cover crop technology in the medium and longer terms.

The Honduran Atlantic coast experience

The following case from Honduras presents a different experience of *Mucuna* adoption and diffusion. The main difference is that the process occurred without the establishment of a farmer organization, like the UNAG/PCaC, to facilitate *campesino* exchanges and workshops. In Honduras, the innovation spread among small-scale hillside farmers. The case prepared by Buckles et al. (1998) examines the use and diffusion of a productivity-enhancing and resource-conserving technology for hillside maize based on the use of the legume *Mucuna derringham*. Box 22 focuses on aspects of adoption/abandonment of the technology, and the farmers' reasons for this.

Box 22 The 'green manure revolution' in Atlantic Honduras

The Department of Atlantida has two agroecological zones: the coastal plain, and the hillsides of the mountain range, Nombre de Dios, that run parallel to the coastline. The hot and humid climate has bimodal rainfall between 2000 and 3000 mm per year and an average temperature of 28°C. The hillside topography is broken, with steep slopes. Small-scale hillside agriculture is characterized by shifting cultivation for the production of maize, beans, manioc, rice, cacao and coffee. Shifting cultivation is typically followed by the establishment of poor quality pastures that quickly degenerate into secondary vegetation. Extensive areas of land are owned by ranchers who exploit only parts of their land-holdings any at one time, and rent out fallow areas to small-scale farmers for maize and bean production, on the condition that they establish pasture after a few years.

The 'abonera' system
The *abonera* system was transferred to the region in the early 1970s by farmers who had migrated from the coastal region of Guatemala and neighbouring Honduran departments. The system consists of a summer *Mucuna* cover crop in rotation with winter maize. Summer maize is planted in separate fields using the conventional technology of shifting cultivation. The *Mucuna* develops a thick mat of foliage during the summer months, reaching the end of its growing cycle by late November. Farmers then cut down the vegetation, and a few

---

weeks later winter maize is sown into the mulch. Winter maize is harvested between March and April. The *Mucuna* is re-established every year, either through natural re-seeding, or by replanting.

The adoption process
The technology spread slowly in the first 10 years following its introduction, and then explosively in the subsequent 10 years. Adoption of the technology increased at a rate of approximately 5 per cent per year, peaking at almost two-thirds of all hillside farmers in the early 1990s. Adoption appears to have levelled off in recent years.

A survey carried out with 128 families in 1992 showed that 34 per cent of the farmers surveyed did not use the *abonera* system in 1992, but many of them had planted maize in a *Mucuna* rotation at some point in the recent past. An estimated 83 per cent of farmers in the study area (over 10 000 hillside farmers) have direct experience with the technology, 16 per cent of whom have discontinued the use of the innovation. Farmers' reasons for discontinuing their use of *Mucuna* were mainly associated with insecurity of access to land. Landowners were more likely to adopt the technology than farmers dependent upon rented land. Land rental arrangements were typically too insecure to justify the establishment of *Mucuna*/maize association by landless farmers. However, farmers with squatters' rights and official land titles were equally disposed to adopt the *abonera* technology.

Adoption rates increased with farm size, as the rotation of *Mucuna* and maize required additional land for the cultivation of summer maize and other crops. However, adoption rates were still relatively high among farmers with less than two hectares. Table 22 presents the survey results and shows the relation between farm size and *Mucuna* adopters.

Table 22 Adoption by farm size classes – landowners only

Farm size (hectares)	With *abonera*		Without *abonera*		Total	
	n	%	n	%	n	%
0–2	5	55.6	4	44.4	9	100
2–5	19	76.0	6	24.0	25	100
5–10	17	70.8	7	29.2	24	100
>10	31	86.1	5	13.9	36	100

Source: Buckles et al. (1998)

Lessons drawn from the Honduran case

- Cover crop technologies can spread without external facilitation.
- The adoption of the cover crop technologies was positively correlated with the size of landholding.
- Landowners were more likely to adopt cover crop technologies than farmers dependent upon rented land.

Conclusions for the potential and limitations for cover crop adoption and diffusion

Diffusion between communities can be enhanced by facilitation by an outside agency. This facilitation might include the organization of exchange visits, workshops and field days. Transport between communities, provision of logistics, and assistance in the use of visualization tools are some of the tasks taken on by external actors. However, the main sources of motivation and stimulation occur from the interaction between the farmers participating in such events. The case study from Nicaragua stresses the importance of exchanges between different communities and the successful adoption/ adaptation of innovations resulting.

Merida Workshop participants identified the limited diffusion of cover crop innovations at village level as a common situation across projects. A shared vision of the participants was to increase the use of cover crops within the communities as well as within the regions of intervention. To understand more about the underlying factors leading to limited adoption/diffusion, a working group carried out a 'force-field' analysis aimed at the identification of enhancing and limiting factors.

The protocol of a 'force-field' analysis is the following:

1 description of the present situation
2 definition of an ideal future situation to be reached
3 identification of factors which either limit or enhance (or both) the advance towards the desired situation.

Figure 9 represents the outcome by the working group.

The factors that can have both positive and negative impacts on the development process towards the future situation are explained in more detail below.

- *Migration*: In Honduras and Nicaragua migration led to the spread of cover crops innovations and contributed to increased diffusion. In Guatemala, in contrast, migration has led to the spread of extensive

Figure 9 Force-field analysis of factors affecting the diffusion of cover crop (CC) technologies

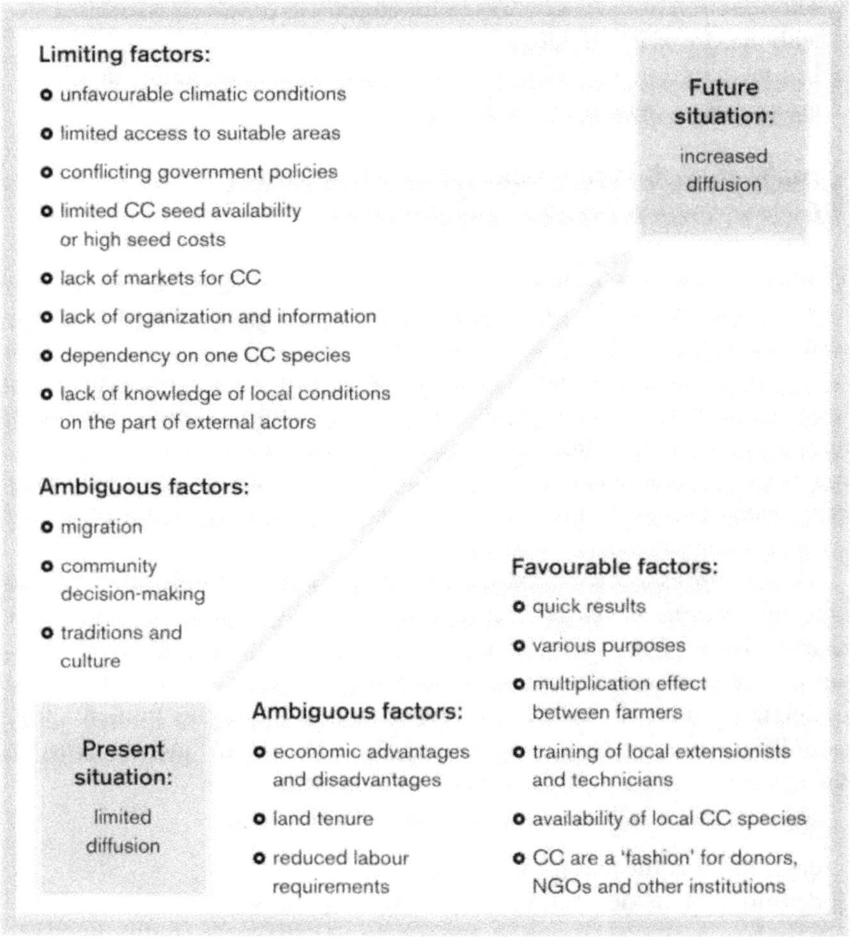

Limiting factors:

- o unfavourable climatic conditions
- o limited access to suitable areas
- o conflicting government policies
- o limited CC seed availability or high seed costs
- o lack of markets for CC
- o lack of organization and information
- o dependency on one CC species
- o lack of knowledge of local conditions on the part of external actors

Future situation:

increased diffusion

Ambiguous factors:

- o migration
- o community decision-making
- o traditions and culture

Favourable factors:

- o quick results
- o various purposes
- o multiplication effect between farmers

Ambiguous factors:

- o economic advantages and disadvantages
- o land tenure
- o reduced labour requirements

- o training of local extensionists and technicians
- o availability of local CC species
- o CC are a 'fashion' for donors, NGOs and other institutions

Present situation:

limited diffusion

livestock-keeping, which favours the establishment of grassland and reduces the diffusion of cover crop innovations.

- o *Traditions and culture*: Knowledge of local cover crop species and the use of local resources can contribute to the diffusion and adoption process of cover crop innovations, as it provides an example to other farmers who do not have this specific tradition. Examples of local cover crop technologies have been provided in Chapter 2 (e.g. Garotilla (*Medicago hispida*) intercropped with oats in Bolivia). On the other hand, traditions like slash-and-burn agriculture might negatively influence efforts made to promote sedentary cover crop innovations. An experience reported from Mexico shows that *Mucuna* displaced a local plant which emerges naturally in association with maize and which is traditionally appreciated as a vegetable. This reduced the interest of local farmers in planting *Mucuna*.

- *Land tenure*: Land tenure is an important factor influencing adoption. If farmers have the legal ownership of their land, or if they have a guaranteed right of permanent use of a certain area (communal land-use agreements), then the use of cover crops is more attractive to them. Farmers who are working on short-term rented land are reluctant to use cover crops, at least for purposes of soil conservation and improvement.

- *Land availability*: Reduced landholding size is an obstacle to the widespread diffusion of cover crops. Farmers living in the humid tropical forest of Nicaragua generally have access to about five hectares of land, which they partly cultivate and partly keep under fallow (Rivas and Zamora, 1996). Under these circumstances cover crops offer a potential for restoring soil fertility more quickly than the natural fallow. For farmers living on the Pacific coast of Nicaragua the situation is different. They have access to less than a hectare of land and they need all the available land for crop production. If cover crops do not contribute directly to family necessities there will be limited chances of their adoption. Similar experiences from Honduras showed that the diffusion of cover crops was difficult in areas where people used their land for intensive horticulture production. They did not see a place for cover crops within their systems (CIDICCO, 1997).

- *Economic factors as incentives or disincentives*: The potential economic benefits of using cover crops constitutes another important aspect influencing cover crop adoption. Farmers who are looking for a direct economic benefit from cover crops are encouraged if a local market exists for the products (seeds) of the cover crops. On the other hand, the absence of markets, or the lack of information on potential markets, reduces the likelihood of adoption.

Local or regional markets for cover crop seeds are not yet established. Seeds are often exchanged on an informal basis between projects or between individuals. Some private companies have tried to establish more formal marketing strategies for cover crops seeds, but prices are high and they are therefore not accessible to small-scale farmers (e.g. a company in Bolivia and Mexico has offered *Mucuna* seed at US$5 per kg).

Experiences from Nicaragua show the reduction of maize production costs by using *Mucuna* due to reduced labour requirements and external input substitution. Labour reduction, as an indirect economic benefit, might or might not be recognized by farmers.

- *Labour requirement*: Some experiences show that farmers perceive a reduction in labour through using cover crop technologies. They report reduced time requirement for weeding and soil preparation as the soil becomes looser due to the incorporation of organic material, and a reduced weed presence. Farmers that have changed from a slash-and-burn system to the use of cover crops report a reduction in labour input. However, other experiences, especially with minimum tillage technologies, have shown that farmers perceived a work input increase associated with these practices.

A series of positive 'forces' considered to favour the diffusion of cover crop technologies were identified during working group discussion.

- *The availability of different cover crop species*: A range of options exist in terms of the choice of cover crop species introduced by farmers. The working group agreed on the importance of including local species in the range of options to be looked at. On the other hand, the number of species selected should not exceed two or three species as this might lead to a less-detailed comparison.
- *A limited number of seeds are enough to get started*: After one season the farmer can obtain enough seeds from the few plants established to increase the area during the following cropping cycle. It is therefore unnecessary for projects to obtain and distribute large amounts of seeds.
- *Cover crops are a new 'fashion'*: As seen in recent projects for rural development and natural resource conservation, financial and technical support at national and international level is directed towards concepts of alternative agricultural technologies. Organizations involved in rural development can make use of this fashion by translating it into more sustainable long-term development concepts. An important aspect is that local cover crop species should be included, because at present the emphasis is on a limited number of introduced species (*Mucuna* and *Canavalia*).

The negative forces identified by the Merida Workshop participants are principally related to policy issues and inappropriate diffusion approaches.

- *Government policy*: Development programmes implemented by local governments are often not compatible with the objectives of non-governmental programmes promoting cover crop technologies. Credit and subsidies are given for external input based agriculture with potential for cash crop production. Moreover, the programmes are frequently short term. In Mexico, for instance, PROCAMPO (a governmental subsidy programme for agriculture), provides subsidies for maize producers whereby the amount received is related to the area of land cleared per year for maize cultivation. This encourages farmers to cut down larger forest areas in order to increase the monetary benefits obtained. On the other hand, NGOs promote permanent land-use practices aimed at increasing production from reduced areas.

Campesino-to-*campesino* diffusion requires an appropriate social environment, where communication and information exchange is possible. The Nicaragua case favours this approach. Limitations have been experienced in areas where highly paternalistic programmes have been implemented.

● *Diffusion and extension approaches:* The working group pointed out that the external actors implementing programmes often lack knowledge of community realities. Problem analysis and identification of potential solutions are insufficient in many cases, reducing the likelihood of successful promotion strategies. Lack of communication, as well as the use of inappropriate communication media between external and local actors, are major reasons for the limited acceptance and diffusion of cover crop technologies. Limitations to diffusion were identified between communities, as well as within communities.

The working group identified the concept of voluntary promoters as a limitation to a programme's sustainability. If the *campesino* does not receive any compensation for time and effort spent in promoting the innovation within his or her community, s/he may lack the motivation and incentive to continue.

Another crucial issue looked at in the working group was the question of the source of information. Where did the innovation come from initially? Was it a local innovation, or was it promoted by an external organization; and if the latter, how was it introduced to a new village? The participants of the working group shared their individual experiences in order to identify different options. The list presented in Table 23 summarizes the results.

Of the 10 projects listed in the table, three mention farmers' local knowledge as the principal source of information on cover crop innovations. However, most projects introduced the cover crop concepts to communities through exchanges with other NGOs, farmers' exchange visits or farmer promoters, as presented in the case studies from Nicaragua and

Table 23 Source of knowledge on legume innovation

Organization	Source of knowledge on legume innovation
Manatlan Project (Mexico)	Local farmers' knowledge of traditional mixed cropping systems and support from World Neighbours (NGO)
CIAT (Bolivia)	Formal research system, working through intermediate users (NGOs and GOs)
Plan Piloto Forestal (Mexico)	Through *campesino* promoters
CENTA (El Salvador)	Secondary information from several NGOs and a workshop
FMVZ -UADY (Mexico)	Exchange visits and rural appraisals
CONSEFORH (Honduras)	Visit to research station
ALA 90 (Paraguay)	Exchange visits and discussions
Costa Norte (Honduras)	Spontaneous movement among farmers
Peten (Guatemala)	NGOs
AGRUCO (Bolivia)	Based on existing local knowledge as point of departure

Honduras. Two of the projects listed in the table (CIAT and CONSEFORH) generated the innovations within their research activities. In the next chapter we present two cases of NARS research strategies for cover crop innovation development.

CHAPTER 6

Research strategies for cover crop innovations

Technological development has been perceived in the past as being the task of research institutions, often working in a 'top-down' way removed from the farmers' reality. This has often resulted in technologies that are either irrelevant or inadequate to the needs of small-scale farmers. The importance of links between the different actors involved in agricultural knowledge systems became obvious when research institutions tried to address resource-poor farmers, especially those in complex, diverse and risk-prone systems, with their off-the-shelf technologies. Adoption rates were usually low because the technologies did not coincide with the perceived needs of farmers, and uptake pathways were inappropriate or non-existent. As a reaction to this shortcoming, research strategies have been readjusted over recent years, leading to a range of different approaches developed to large extent from, or as a response to, farming systems research. On-farm research and farmers' experimentation methods have arisen, aimed at improving information flow between actors and understanding the complexity of LEIA systems.

In the following pages two cases are presented of NARES which are involved in the development of cover crop innovations for LEIA systems. The cases show different approaches in terms of their research strategies, farmers' participation and uptake pathways. It is important to note that these two case studies are not meant to present a recipe for ideal linkage mechanisms between actors, but rather aim to provide a basis for the analysis of research strategies, and to stimulate further discussion and ideas. Each approach has its specific potentials, strengths and limitations, from which basis alternatives could be developed.

Applied and Adaptive Research at the Centro de Investigación Agrícola Tropical (CIAT), Bolivia

CIAT is a regional research centre covering the Department of Santa Cruz in eastern Bolivia. Its mission is to: *contribute to sustainable development in the Santa Cruz Department through agricultural and forestry technology.* Its objectives are to generate, validate and transfer technology, and provide training in the use of this technology and methods for its transfer. CIAT's clients are NGOs, regional development projects and programmes, farmers' associations and local government (municipalities).

Research approaches

CIAT works though the implementation of projects in the three main production systems found in Santa Cruz Department: mechanized agriculture, extensive ranching and slash-and-burn (Linzer et al., 1997). Serious soil degradation and weed management problems are common to both mechanized and slash-and-burn agriculture. Figure 10 shows the relationship between *technology generation* (conventionally in its Experimental Station), *selection* of technology (regional research centres), *validation* (on farmers' fields) and *transfer and training* for intermediate and end-users of the technology (by the Training and Transfer Co-ordination Unit). The words in italics are translations from CIAT's own publicity material. However, it might be more appropriate to describe the work at the Experimental Station as researcher selection and adjustment of technology under controlled conditions; that at the regional research centres as the testing by researchers of technologies thought to be suitable for regional conditions; and the work on farmers' fields as the joint testing, adjustment and selection of a range of options by farmers, researchers and extensionists. Further adaptive research is continued by farmers, unsupported as yet by CIAT.

Figure 10 Technology development and diffusion by CIAT

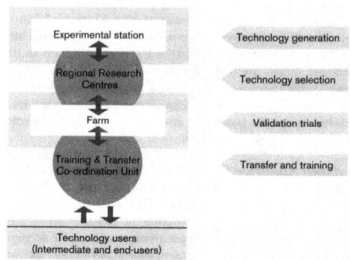

At the level of CIAT as an institution, the process of research and transfer is linear. However, some of its projects are now using a more fluid approach to research with farmers. One such is the Sara Ichilo Adaptive Research Project, which is a collaboration between CIAT and NRI (UK), and which is described below.

Figure 11 Cover crop research and dissemination activities at CIAT

Experimental Station
Researcher screening of cover crop species and husbandry for their potential introduction into specific regions and/or systems

Regional Research Centres
Evaluation of cover crop species under physical conditions that are broadly representative of the region, and within simple potential crop associations

Farmers' fields
(a) Joint farmer, extensionist, researcher evaluation of covers in a range of crop associations and livestock applications through 'collaborative' on-farm research
(b) Researcher-managed trials to clarify or further develop aspects arising from collaborative evaluations ('contract research')

Dissemination
(a) Through on-farm collaboration of farmers and extensionists with researchers
(b) Through production of dissemination materials
(c) Through workshops, training and consultancies

While *ad hoc* linkages – based on personal interest – exist between these four components, and periodic workshops bring together those involved, there is no explicit co-ordination mechanism to promote common cover crop strategies that are responsive to farmers' opportunities and constraints.

Research at the Experimental Station and regional research centres has focused mainly on the selection of varieties. From 56 introduced materials, 28 have been identified as having good adaptation to specific conditions in Santa Cruz Department. Selection has been on yield, nutrient content, ability to withstand stress (e.g. drought or flooding), and the ability to control weeds. Detailed results of this work can be found in Linzer et al., (1997).

At the farmer level is the CIAT/NRI Sara Ichilo Adaptive Research project, whose purpose is to:

> develop and promote technologies and research methods that enable the evolution of sustainable livelihoods for rural families in the provinces of Sara and Ichilo.

This includes a substantial cover crop component (CIAT/NRI, 1997). The situation confronted by the project is that migration of populations into the tropical humid lowlands of Bolivia has resulted in the degradation of soil and forest resources. Smallholder farming families have limited access to technological options or income-generating opportunities to replace their unsustainable slash-and-burn practices, which use the fertility of the forest biomass as natural capital. Migratory agriculture leading to the extension of the agricultural frontier is encouraged by the requirements for land from mechanized agriculture and ranching. The problems faced are categorized between 'international' and 'local' problems in Box 23.

Box 23 Problem scenario for smallholder agricultural research in lowland Bolivia

International problems

- *Tropical moist forest loss* (loss of biodiversity and habitat, reduced carbon sequestration, reduced water quality and flow, livelihoods of indigenous forest dwellers affected)
- *Natural resource degradation* (soil degradation, water quality, air quality through burning)
- *Poverty* due to unsustainable and unprofitable agricultural practices in the tropical lowlands of Central/South America.

Local problems

- *Loss of nutrients* through burning and erosion
- *Short cropping periods* and reducing fallow periods
- Rapid development of *aggressive weed flora* following forest clearance
- *Lack of resources* for farm inputs/improvements (financial/human)
- *Lack of income*-generating opportunities
- *Underdeveloped* infrastructure, local institutions, input supply and markets.
- *Short-term demand* for food/income.

The issues are extremely complex, with farming households under different circumstances opting for different routes through the capitalization/stabilization process (Richards, 1997; Wachholtz, 1997). Each stage has its own characteristics, and specific requirements for technical/institutional support (Muchagata, 1997).

Adaptive research elements

The research approach adopted by the CIAT/NRI Sara Ichilo Adaptive Research Project is one of participatory on-farm research, supplemented by researcher-managed trials to provide a large 'basket of options' from which farmers can chose. However, it is understood that addressing the complex situation requires the development and testing of a strategy, rather than the identification and validation of technology (see Conclusions and future strategies).

In order to make a significant impact in the project area, the implementation of the on-farm trials programme is shared between CIAT and local governmental and non-governmental organizations. This collaboration arose from recognition of the need for co-ordination of adaptive research in the area, following a survey carried out by the project on the agricultural research and development activities of institutions within the project area.

CIAT has implemented a substantial on-farm research programme through the creation of a number of decentralized zonal research and technology transfer teams, and by supporting the piloting of participatory research methods.

Working directly with over 200 farmers has required the establishment of computer databases to record details of each farmer, their trials and results, and spreadsheet programmes to analyse quantitative economic and agronomic data. The wide range of farmer circumstances in the project area has been categorized using a Recommendation Domain classification. This has provided a framework for farmer-based studies, the interpretation of information from on-farm research, and the formulation of recommendations for intermediate and end-users of the project's outputs.

The on-farm trials incorporate a number of common principles:

- *Diversity*: providing multiple income sources and efficient use of labour, soil, water and sunlight.
- *Soil cover*: for protection of the soil from erosion and high soil temperatures, control of weeds and maintenance of soil fertility and structure.
- *Sequences* towards stable farming systems based on perennial species.
- *Integration* of annual crops, perennial crops, forestry species and livestock activities in one programme.
- *Simultaneous research and development*, with involvement of NGO, grassroots and government extension agencies in the on-farm research programme.

The cover crop elements of the annual, perennial and agroforestry systems under test include a range of species that are appropriate to different associations or agricultural systems development stages, when growth habit and aggressiveness, longevity, rooting system, shade, drought and temperature tolerance are all considered. Promising species are: *Mucuna* spp. for maize-based cropping systems, pasture regeneration and perennial fruit orchards; *Pueraria phaseoloides* (tropical kudzu) for pasture regeneration and perennial fruit orchards; *Calopogonium mucunoides* for relay cropping or rotations with rice and for perennial fruit orchards; *Arachis pintoi* (forage groundnut) for perennial fruit orchards (although farmers have reported problems of slow establishment and poor competitiveness with weeds); *Canavalia ensiformis*, a versatile and easily managed annual legume for rotations with annual crops, and ground cover in perennial crops; *Cajanus cajan* (pigeon pea) and *Crotalaria* spp. for intercropping with annual crops and early stages of perennial orchards; *Desmodium ovalifolium*, a shade-tolerant perennial legume for later stages of perennial plantations.

Researcher-managed trials with a cover crop component are investigating the following:

- the intercropping, relay cropping or rotation of *Calopogonium mucunoides* in rice in order to extend the period of cropping between bush-fallows
- the selection of varieties and management practices for winter cover crops that are tolerant of cool temperatures and moisture stress
- the identification of potential cover crops for association with the cash crop *Bactris gasipaes* (peach palm), in order to reduce labour requirements and establishment costs.

These trials are carried out on rented farmers' fields that have the same fertility and weed constraints faced by farmers. Farmers are invited by researchers to evaluate these trials and to suggest improvements.

A number of participatory fora are used for the evaluation of technologies.

- *Workshops* have been held for joint evaluation of the programme by researchers, field technicians and farmers. The methods used have varied from technician-driven question and answer sessions, to the use of farmer group discussions from which technicians were excluded until farmers presented their conclusions. These workshops have been very useful, together with *individual on-farm interviews* carried out when project technicians visit the trials routinely, to identify suggestions for modifications to the on-farm programme and topics for formal researcher-managed trials.
- *Field days* are held twice a year at the site of on-farm trials as a forum for interchange between the farmers responsible for the trial, technicians, and the community in which they live. They also provide valuable feedback to researchers.
- *Surveys.* A series of surveys have been undertaken. One has looked at a sample of the 89 communities covered in order to assess adoption and adaptation of the technology by those farmers who have collaborated directly with the project and those who have not (Warren, 1997). Some conclusions are presented below. A second survey of 110 collaborating

farmers looked at the reasons behind changes made by farmers to on-farm trial components during their implementation, and also examined the uptake of technologies by collaborating farmers. This has been significant, as shown in Table 24. A third survey looked at the 'fit' between the economic circumstances of farmers and the technologies being evaluated. All surveys used semi-structured interviews as their main tool, complemented by a number of other participatory rural appraisal methods to ensure triangulation of information. The software package SPSS (Statistical Package for the Social Sciences) was used for survey analysis.

Table 24 Uptake of cropping systems involving cover crops by farmers collaborating in on-farm trials in Bolivia

Cropping system	No answer	Have adopted	Would like to adopt	Not interested in adoption	Don't know
Bananas plus covers	–	50%	50%	–	–
Rice plus covers	–	28.6%	57.1%	14.3%	–
Citrus with covers	7.7%	53.8%	33.3%	–	5.1%

Results of a survey of 110 farmers.

Lessons learned

The research, which was conducted with individual collaborating farmers and supervised by technicians from CIAT and local NGOs, has resulted in a great deal of useful quantitative (agronomic and economic) and qualitative information. However, there are also lessons to be learned, and improvements that could be made to the methodology.

- Haste in trying to establish a large number of on-farm trials in a short time resulted in reduced participation of farmers in the planning and trial design process.
- The programme focused on individual farmers and not on communities. This resulted in other members of the community being mostly unaware of the trial's objectives and results.
- There was a lack of clarity in setting the objectives of the on-farm trial programme (i.e. whether the objective was to gather clean sets of

statistically viable quantitative data, or alternatively to allow farmers to manage the trials in their own way). This led to a compromise situation in which neither was fully achieved.

An adoption study (Warren, 1997), in addition to providing important information about the adoption or non-adoption of specific technologies (see Chapter 5), also provided useful feedback to the research and development process, including the need to:

- improve the level of awareness of the project and its activities among non-collaborating farmers
- develop links between communities and credit institutions
- improve the availability of seeds and planting materials
- improve the project's knowledge of the activities and priorities of smallholder farmers in the project area.

Uptake pathways

The technical experience of the project was distilled into a series of dissemination materials (bulletins, posters, manuals and videos) of direct use to intermediate users (governmental and NGO dissemination agencies) and to some final end-users (farmers).

Conclusions

The strategy for developing and promoting sustainable agriculture in forest margins that is emerging has a number of components:

- the development and delivery to farmers and dissemination agencies of a wide range of technical alternatives to aggressive slash-and-burn techniques through participatory research, linked closely to 'conventional' research
- the identification of income-generating opportunities, and the development of support services (inputs, advice, credit, marketing) necessary to enable these to be adopted
- the evolution of appropriately structured and resourced formal and community-based institutions
- close liaison between research and development programmes and institutions, to achieve common goals jointly and ensure uptake of research outputs
- the definition and implementation of a policy environment conducive to sustainable land use.

Within this strategy cover crops have a key role to play, and the complementary applied and adaptive research at CIAT has, with farmers' collaboration, achieved considerable advances in identifying suitable cover crop species, husbandry methods and system niches.

Action–research with campesino farmers in south-east Mexico

The Agricultural Extension and Systems Research Department (DISE) of FMVZ-UADY (Facultad de Medicina Veterinaria y Zootecnia, Universidad Autonóma de Yucatán) is involved in action–research with *campesino* communities and NGO partners in south-east Mexico. DISE addresses the problems of *campesino* agriculture by providing an interface between FMVZ-UADY and *campesino* communities. This was necessary because FMVZ-UADY had previously, in common with other state-run agricultural institutions, focused only on commercial agriculture. Questions of food security and *campesino* family well-being had not been included on the research and training agendas of the official agricultural institutions. Subsistence and semi-commercial *campesino* agriculture in south-east Mexico has been supported by small-scale NGOs in terms of extension and technological innovation, most often as a means of bringing about forest habitat conservation.

The DISE is working in south-east Mexico, mainly in the Yucatan Peninsula. Most of the participating *campesino* families are Mayan; others are migrant families of different ethnic groups. The action–research sites are indicated in the map (Figure 12).

Figure 12 Location of DISE action–research sites in south-east Mexico

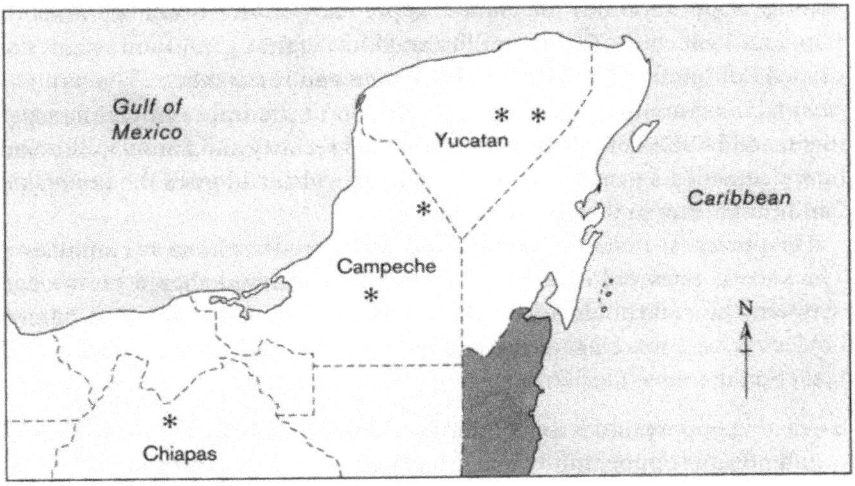

DISE's main objectives

The overall objective of DISE is to facilitate endogenous change within *campesino* communities to improve food security and family well-being through a sustainable use of natural resources. The objective has been derived through working closely with different indigenous communities in south-east Mexico, and has been formulated as a response, by a natural resources research institution, to the needs of people in the poorer agricultural sectors. The objective is achieved through two main strategies:

- *action–research*: facilitating processes of *campesino* experimentation, providing technical assistance to *campesino* families, generating relevant scientific information.
- *building uptake pathways*: promoting the use of participatory methods of appraisal and technology development by local NGOs and *campesino* organizations, promoting *campesino*-to-*campesino* interchanges; providing technical assistance to community development projects.

DISE provides an interface between the Faculty and *campesino* communities. A two-way flow of information has been facilitated whereby *campesino* innovations are supported by technical information generated from Faculty research. As a result of this information flow, the Faculty's research has been oriented towards agricultural problems as perceived by *campesino* families.

Development of an action–research methodology by DISE

The methodology that has been developed by DISE for achieving action–research linkage with *campesino* communities can be summarized in the iterative application of four phases: Appraisal, Convergence, Experimentation, and Reflection. These are illustrated in Figure 13. All four phases are carried out jointly between farming families and researchers. The nature of *campesino* communities, the gender division of natural resource management, and DISE's objective of improved food security and family well-being have demanded a gender-sensitive approach which addresses the *campesino* family as partner in the research process.

The process is iterative and the impact of the four phases is cumulative. The success achieved is defined in terms of increased shared knowledge between the stakeholders involved and the degree to which the agreed objectives are met. During the action–research and linkage process DISE takes on the role of facilitation by:

- creating opportunities to identify, with the *campesinos*, the constraints, and possible opportunities and solutions, within their farming system

- providing access to external knowledge, and exchanges between local and external knowledge
- encouraging the *campesinos* to take control of the experimentation and innovation according to their needs and priorities
- providing tools for the *campesinos* to develop a process for the analysis and systematization of the findings of the experimentation phase.

Figure 13 The phases in the iterative action–research process

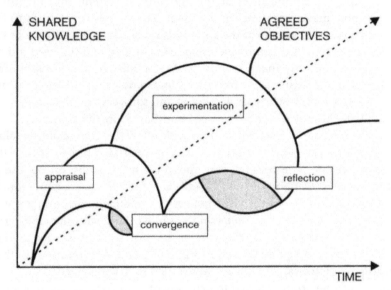

Cover crop innovation development and diffusion

During participatory appraisals with several *campesino* communities, the need to improve backyard animal production was expressed by participants as a way of improving both food security and family well-being. This became the initial focus of work. Aspects of preventive health measures for backyard livestock, including zoonoses, conservation of *criollo* and indigenous livestock breeds and alternative livestock feed/forage sources, were included from the earliest phases of the work. It was quickly learned that the integration of backyard production (a female domain) and the traditional *milpa* (a male domain) had been the cornerstone of *campesino* well-being and had to be understood and built upon to improve food security and livelihoods. Research on participatory innovation development within traditional slash-and-burn maize cultivation contributed to the widening of DISE's focus on crop/livestock interactions in *campesino* systems, including action–research on cover crop technologies.

In the following account, the process of integrating crop and livestock husbandry into a common research strategy is explained. The activities of the four phases of the action–research model are explained.

In 1994 a research project was initiated with rural communities, in order to explore together with the *campesino* families alternatives to improve livelihoods based on local resources. There was a lack of information on the concepts and the strategies for *campesino* livelihood maintenance so the research was initiated through a series of rapid rural appraisals. This gave insights into the complexities of the agricultural systems and identified areas for potential innovation development. This phase corresponds to the *appraisal* phase. The results of the appraisals revealed the importance of the *milpa* and the homegarden for the *campesino* families, and the need to find alternatives to improve the productivity of these subsystems. The appraisals also revealed the existence of important local knowledge of the environment (classification systems for vegetation and soils, ceremonies, etc.).

The next phase was one of *convergence*. Based on the findings of the appraisals, the research team discussed with interested *campesinos* possible management strategies to improve *milpa* productivity and to reduce the deforestation rate. The research team provided information on the use of cover crops (*Mucuna pruriens* and *Canavalia ensiformis*) as green manure in maize production. Exchange visits between *campesinos* collaborating with the researchers and another *campesino* group working with the cover crop system were organized. During the visits not only was information exchanged but also legume seeds were obtained. At this point, the objectives of the *campesino*/researcher linkage were established and agreed.

The *campesino* groups started to experiment with cover crops on a small scale in their own agricultural system. They modified the new system according to their specific needs, which resulted in variations in the cropping pattern, the species and varieties planted, the mulch management, the sowing and harvest dates, and the use of the crops obtained. This is termed the *experimentation* phase. Important mechanisms to ensure a joint learning process were the establishment of focus groups for discussion and field observations, workshops with participating *campesinos* from various communities, and further exchange visits.

Results were compared between the participants, and the *campesinos* established a list of criteria to evaluate the modifications made. These mechanisms provided a phase of *reflection*. During the reflection phase the *campesino* vision of the *milpa*/backyard animal production integration was brought to the fore and the next stage of the action–research project was dedicated to exploring the new possibilities of enhancing this integration afforded by the innovations in the crop system (Gündel, 1999).

Linkages between on-station research and community-based action–research

The dual strategies of DISE (action–research and building uptake pathways) are achieved through three main activities, shown in Figure 14. These activities run in parallel, and are:

- generating relevant scientific information
- facilitation of *campesino* experimentation
- promoting *campesino*-to-*campesino* interchanges.

Figure 14 shows the interrelation between these activities and gives details of how they have been carried out over the last couple of years with regard to cover crop innovations in staple crop production and alternative livestock feeding strategies.

Figure 14 The interrelations between DISE action–research activities

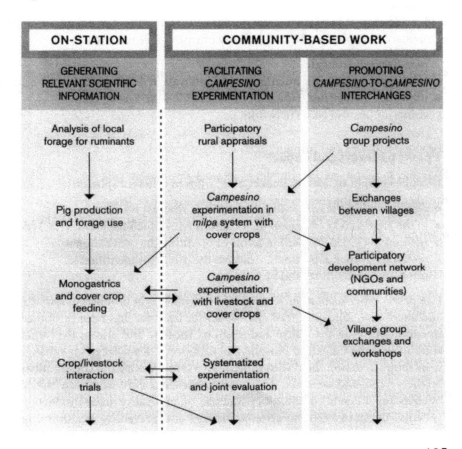

Both strategies – action–research and building uptake pathways – are pursued by all researchers of DISE, albeit to different extents. The distinction between them depends upon the stage the work has reached. Action–research leads to the building of uptake pathways due to the engaged nature of the work in the *campesino* communities and the contacts established that are utilized for diffusion. All three activity axes are important for each strategy. However, action–research involves mostly activities of *campesino* experimentation and the generation of appropriate scientific knowledge, while the building of uptake pathways involves both *campesino* experimentation and promoting *campesino-to-campesino* interchanges.

The expected outputs of the cover crop action–research are:

- development of alternative cropping strategies, based on permanent land use
- improved feed availability for homegarden animal production
- improved crop/livestock integration through multi-purpose cover crop technologies
- identification of local cover crop species and management practices
- increased economic and social viability of sustainable resource management strategies
- local capacity for experimentation and innovation development
- improved linkage mechanism between FMVZ-UADY, *campesino* communities and NGO networks.

Uptake pathways and impact

DISE has sought to achieve impact of its work in different spheres:

- directly within the *campesino* communities where it is active
- in the NGOs and *campesino* organizations with which the group is linked
- within FMVZ-UADY itself in terms of orienting the conventional research of other departments towards the problems identified in *campesino* agriculture, and by training.

Campesino experimentation, looking into innovations in production systems, and alternative sources of feed and fodder for livestock, has been promoted in communities in the centre and south of Yucatan, and also in the buffer zone around the Calakmul Biosphere in Campeche. Knowledge interchange workshops have been held with more than a dozen *campesino* communities on these topics and *campesino-to-campesino* exchange visits have been carried out between different communities. Manuals on innovative maize production systems, alternative feeding systems for pigs and preventive healthcare of backyard poultry have resulted from these activities. Different groups are currently experimenting with the use of cover crops in maize systems and feeding cover crop grain and forage to backyard pigs.

Since 1994, links have been established with regional community development and conservation NGOs and with innovative *campesino* groups in communities linked by a participatory development network. These connections represent effective uptake pathways for the technology, information and knowledge generated by DISE's activities and also by other departments of FMVZ-UADY which produce technology appropriate for *campesino* families.

CHAPTER 7

Pulling the concepts together

In this final chapter we intend to discuss the lessons we have learnt from the preceding thematic chapters. First we summarize the main functions and purposes of cover crops developed in the previous chapters. This will provide us with the basis for a holistic cover crops definition. We shall then develop a conceptual framework to explain the diversity of functions and purposes cover crops may have in different systems and at the different stages of system development. In the second part of this chapter we shall look at the main conclusions of the Merida Workshop so as to identify the potential for, and the constraints on, future cover crop adoption and adaptation. The general conclusions from the workshop will be translated into strategies for the use, promotion and research of cover crops that might be taken up by different actors.

Towards a holistic cover crop definition

By applying a function/purpose perspective and drawing together the information presented in the previous chapters we can arrive at conceptual and holistic definitions of cover crops that take into account their functions (bio-physical processes and interactions that cover crops are involved in) and purposes (the uses to which cover crops are put by farmers taking advantage of cover crops functions) within agroecosystems.

Figure 15 illustrates a series of functions that cover crops can perform within agroecosystems. These functions operate at field and farm levels. Particular functions have greatest relevance to one or other of the levels.

Among the functions that cover crops perform at the field level are: production of organic matter, recycling of nutrient, allelopathy and competitive growth, provision of soil cover, nitrogen-fixation – in the case of leguminous species – and insect suppression in some cases. At the farm level the functions of cover crops are mainly the production of grain and forages.

Cover crops contribute to their environment in different ways that can lead to a reduction in soil erosion, maintenance of a positive water balance and an enhanced nutrient availability. These functions occur through:

- providing foliage cover which protects the soil from the elements
- the creation of a micro-climate whereby soil water evaporation is reduced
- the transformation and storage of soil nutrients as organic matter, which are released once the plant senesces.

These contributory functions benefit the other crops grown in association with cover crops.

We will now derive a *functional* definition of cover crops:

> A cover crop is a live soil-surface cover used as a temporal or spatial component in annual or perennial cropping and agroforestry systems which performs one or more functions such as organic matter production, nutrient recycling, allelopathy and/or competitive growth, soil cover, nitrogen-fixation, and insect suppression, while producing forage and grain.

Figure 15 Functions of cover crops within agroecosystems

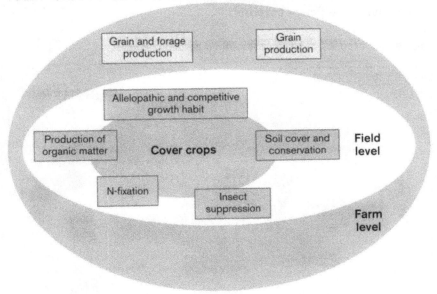

Now we shall transform the functional definition into one that takes into account the purposes that different stakeholders may have for incorporating cover crops into existing agroecosystems. In doing this we shall add the beyond-the-farm level to the model. This level might be the watershed, the region, or national and even international levels.

The functions that cover crops perform can be utilized for different purposes at different levels. For example, at the field level the functions of organic matter production, nitrogen-fixation and soil cover provision can be used in soil fertility management, soil structure improvement and soil and water conservation, while the soil cover, allelopathic and competitive growth characteristics can be utilized for weed control. The production of grain and forages at the field level can be used for the farm-level purposes of providing animal feed and/or human food production, income generation and crop diversification. The priority given to these various purposes depends upon the objectives and needs of the different users, and hence will vary among stakeholders.

The range of functions that different cover crops have allow stakeholders to identify those that are relevant to their circumstances and which fulfil their objectives. Hence the purpose(s) for the incorporation of a given cover crop in an agroecosystem is that it performs one or several functions deemed desirable by one or several stakeholders. Figure 16 illustrates the different purposes that cover crops fulfil at different levels. By comparing Figures 15 and 16 we can see how the different cover crop functions can be used to fulfil the different purposes.

Figure 16 Purposes of cover crops within agroecosystems

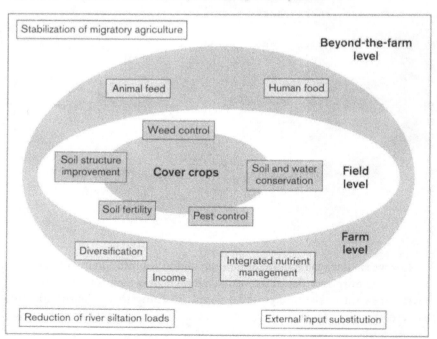

Thus we arrive at a purpose-oriented definition of cover crops:

> A cover crop is a live soil-surface cover used as a temporal or spatial component in annual or perennial cropping and agroforestry systems for fulfilling one, or several, purposes including weed management, soil productivity restoration and maintenance, the provision of livestock feed and/or human food, and diversification of income generation options.

The range of stakeholder purposes

From the preceding discussion it may be appreciated that cover crops can perform different functions in agroecosystems and that these are utilized for different and multiple purposes depending on situation and need. For example, cover crops may contribute to farmers' livelihoods but they may also be promoted by external stakeholders for purposes at beyond-the-farm level. To improve our conceptual understanding of cover crops we need to explore further the idea of different stakeholders' purposes. We shall now expand on the nature of the purposes that farmers may have for incorporating cover crops into their agroecosystems.

Farmers' purposes for the inclusion of cover crops in their farming system can be categorized as immediate/recurrent, medium-term, or emergent. Immediate/recurrent purposes are those that are necessary for the farmer, and her or his family, to fulfil in order to maintain their livelihood and farming activities through seasons. Medium-term purposes are those that sustain livelihoods and the farming system over years and longer cycles. Emergent purposes are those that arise as a response to changes as the system evolves, and may be opportunistic or of a systems-repair nature. The different functions that cover crops perform can be utilized for the fulfilment of different purposes as the farming system evolves and develops through processes of intensification and integration (or degrade through resource depletion and disintegration). The different functions that cover crops perform will justify, or not, their choice for inclusion in the system. For example, a single cover crop may be incorporated in a crop association for its weed control ability, and be maintained in the system, once weeds are controlled, for the purpose of providing food/feed grain or soil improvement, or alternatively it might be replaced by another (cover) crop that better fulfils these purposes. The framework in Table 25 illustrates this, using an example of *Mucuna* and *Canavalia* inclusion in a crop/livestock system.

Table 25 Framework to illustrate the relationship between different functions and purposes that different cover crops may fulfil at different junctures in a farming system: *Mucuna* and *Canavalia*

Functions	Immediate/Recurrent	Purposes* Medium-Term	Emergent
Competitive and/or allelopathic growth habit	*Mucuna* – for weed control due to its very vigorous growth	*Mucuna* – may be replaced by other crops, e.g. *Canavalia*, as weed control is achieved	*Mucuna*, to solve an emergent weed problem
Production of organic matter and provision of soil cover	*Mucuna* – rapid and ample cover required	*Canavalia* – to provide leaf material for mulch	*Mucuna* for reduction of erosion during high-risk periods
Production of foliage and grain	*Mucuna* – grain and forage for livestock, other bean species for human food	Other crops including *Canavalia* to provide grain for ruminants	*Mucuna* associated with Gramineae for rapid and abundant forage production

* See text for definition of different categories of purpose

External stakeholders' perspectives on cover crop use

In the previous section, we discussed the purposes that farmers may have in using cover crops. The scope of consideration is now widened to include external stakeholders – that is to say non-farmers. In this wider context (beyond-the-farm level) the utilization of field and/or farm level functions, once aggregated over watersheds or regions, can contribute to the fulfilment of purposes that might include the reduction of siltation load of rivers, external input substitution (fertilizer, herbicides, livestock concentrates), and contributions to the stabilization of migratory farming and unsustainable slash-and-burn systems.

Table 26 shows how the purposes for utilizing cover crops may vary among different stakeholders. Given a) that cover crops vary in their ability to fulfil the different functions and purposes, b) that among the different stakeholders there will been different priority rankings of the purposes, and c) that some of the purposes will be fulfilled more completely by crops other than cover crops, it is possible to understand how little convergence of interests might exist among stakeholders over the utilization of cover crops.

When farmers' interest in using cover crops is weak it may be necessary for external stakeholders to promote the use of cover crops through the demonstration of favourable and complementary functions, or the introduction of incentives and/or subsidies. Indeed the very real benefits of external input substitution at the farm level might become apparent to farmers only once this potential is demonstrated to them. Added to this is the farmers' stake in the beneficial impact of cover crops at the beyond-the-farm level, which might also have to be explored with them. Here there are issues of quantitative changes leading to qualitative change: as more and more farmers in a watershed adopt cover crop use, so a point may be reached whereby the accumulated impact is favourable to everybody in the area, for example through the reduction of siltation problems or the establishment of a viable local market for cover crop products.

Table 26 Cover crop purposes and different stakeholder motives for utilization/promotion

| Cover crop purposes | Stakeholders' motives for utilization/promotion | | | |
	Farmers	Agricultural policy makers	Natural resource conservationists	Rural development agencies
Weed control	reduced labour inputs	import substitution	reduced herbicide use	reduced toxicity problems
Nitrogen-fixation	increased yields	import substitution	reduced inorganic inputs	improved food security
Soil and water conservation	reduced soil (nutrient) losses	sustain productivity	reduced erosion; reduced siltation	improved food security
Pest control	reduced yield losses	import substitution	reduced contamination	reduced toxicity problems
Soil structure improvement	increased yields; ease of soil cultivation	sustain productivity; maintain water quality and quantity	reduced erosion and soil degradation	improved food security
Human food	consumption; income	stimulate local markets		improved food security
Animal feed	increased options	import substitution		improved food security

Pulling it together: multiple stakeholders' perspectives related to specific farming systems and projects

This section presents the conclusions reached by the Merida Workshop participants regarding the current situation of cover crop use in different systems, and the diversity of prevalent purposes. The participants – *campesino* promoters, farmers, extension workers and researchers – reached a set of conclusions which represent an important commentary on cover crops and the opportunities and constraints associated with their wider use. The conclusions are taken from the working group sessions. Some are presented under the title of the themes discussed, while others have been grouped under the title of 'Strategies for improving planning, evaluation and knowledge/information-sharing'. The workshop also identified some research needs that represent challenges to applied researchers. These are listed below.

Cover crops in annual and perennial systems

- Poor seed availability is an important factor limiting the introduction of more diversified cover crops into existing farming systems. There is no organized infrastructure for cover crop seed distribution and technical problems exist in terms of seed production and storage.
- At present the demand for cover crop seeds is increasing. In some regions the availability of cover crop seeds is sufficient, in others not. Mechanisms are required that ensure the availability of new and local seeds for research and extension.
- The establishment and maintenance of perennial crops implies a capital investment. Few *campesinos* are able to leave an area of land unproductive.
- The most important function of cover crops in perennial crops is allelopathic and/or competitive growth, with the purpose of reducing weeding costs.
- There are experiences at the field level of the integration of cover crops with a range of perennial crops under different conditions. However, the range of cover crop species used is limited and the need exists to explore a wider range of cover crops.

Cover crops and soil and water conservation

- Cover crops do not present a single strategy for soil and water conservation. It is important to recognize the complementarity between the use of cover crops and other conservation measures:
  o The contributions of cover crops to the reduction of soil erosion are the following: organic matter, soil cover, increased infiltration, reduced impact of rains.

o On steep slopes, cover crops on their own are not sufficient to reduce soil erosion. Additional measures, such as live barriers, terraces and contour bunds, are required. The lessons learnt from the terrible experiences during and after the passage of Hurricane Mitch through Central America in 1998 serve to underline this point dramatically.
o Factors important in the balance between the conservation of nutrients (decreased leaching, nitrogen-fixation) and crop requirements are: synchronization of crop and cover crop establishment, positive and negative crop interactions, and pests and diseases.

- Cover crops can contribute to improved water quantity and quality by reducing herbicide use, acting as a fertilizer, improving filtration, and reducing sediments. However, the question of who pays for these benefits and in which way remains important. Communities close to water reservoirs lend their services to the consumers. To do so needs incentives and awareness-raising on conservation issues.

- It is important to negotiate a increased convergence between the objectives of *campesino* producers and those of conservation projects. *Campesinos* want to improve their quality of life, increase productivity, and improve their incomes, whereas conservation projects want to protect the environment. Hence, it is important to understand the priorities of the different actors involved. It is necessary to resolve existing problems in political and legal terms regarding land rights, restrictions of use, taxation, prices, and other related pressures to stimulate cover crop use.

Cover crops and livestock components

- Livestock components can be crucial for improving bio-economic efficiency, income generation and human nutrition. The integration of cover crops represents an opportunity to strengthen this role.
- In commercial livestock systems cover crops have the potential to:
  o regenerate pasture land degraded by extensive livestock keeping
  o improve the use of fibre-rich diets through providing good quality forage and grains for ruminants and non-ruminants
  o contribute towards sustainable intensification of small-scale farming systems.
- At present a selection of cover crops for feed uses (grain and forage) are undergoing an evaluation process through *campesino* experimentation and formal research. However, the range of species to be considered has to be broadened, taking into account *campesino* innovations and priorities.
- In some countries the analysis of socio-economic aspects of cover crop/livestock integration has been initiated.

- At the forest/agriculture interface an environmental impact assessment of cover crop/livestock integration is required, looking into implications for natural resource conservation and biodiversity maintenance.

Diffusion and adaptation of cover crop technologies

- The majority of the factors identified that influence adoption of cover crops are non-technical factors. This implies that interventions that include cover crop technology should not only look at technical issues, but also have to take into account a series of other socio-economic aspects, such as gender, and access to inputs, financial resources and technical assistance.
- It is important to consider the wider environment in order to determine positive and negative factors that influence adoption and diffusion. Migration, for instance, could have a negative impact, if *campesinos* bring with them slash-and-burn agricultural traditions, whereas it could also be positive, if those that arrive have knowledge (and seeds) of appropriate cover crops.
- Cover crops are promoted within a wide range of geopolitical situations and agroecosystems. Within these different situations a range of different *campesino* objectives have been identified to explain why cover crops were adopted.
- Based on the experiences of the workshop participants, it was concluded that external objectives and local objectives for cover crop technology utilization seldom coincide.
- Wider diffusion of cover crop technology could be achieved if we knew and understood *campesino* objectives better. This would enable us jointly to develop appropriate strategies based on *campesino* expectations, knowledge and criteria.
- Target groups for cover crop promotion and diffusion have to be identified. Target group definition has to include social, economic and ecological characteristics. Gender differentiation has been neglected so far in most cover crop projects, as have landless farmers.

Strategies for improving planning, evaluation and knowledge/information sharing

- Any kind of project planning should make use of local knowledge as a starting point, taking into account that it varies between communities as well as between *campesinos*, and that it can easily be weakened under the influence of external factors. It is also very dynamic over time.

- In order to value *campesino* knowledge, the attitude of the technicians, including the institution they belong to, is a fundamental determinant of knowledge sharing. Exchange workshops and case studies carried out within training institutions could be useful to facilitate an exchange of existing knowledge and to identify complementarity between external and local knowledge.
- In most cases the methodologies used so far have not allowed researchers to engage with *campesino* community realities. This has led to extractive and inappropriate research. The institutions acting as links between communities and research institutions have to increase their capacity to suggest research topics and their negotiation power to establish contacts between research institutions and communities, which would serve to define topics, products and cost/benefit distribution.
- Projects often fail to evaluate on a systematic basis the impacts of cover crop introduction, apart from the direct adoption response by *campesinos*. Specifically, there is a lack of economic analysis and of assessment of impacts beyond the plot level. It is important to systematize the experiences from Latin America in relation to soil, climate, altitude, rainfall and socio-economic conditions, in order to facilitate the identification of suitable cover crops for specific situations.
- A lack of communication is limiting advances on the basis of previous experience. The need exists to develop further appropriate methods and instruments to improve the links for information exchange between external actors and locals. Sharing information and valuing local knowledge requires a financial and organizational infrastructure. It is important to support organizations like CIDICCO, and to work towards an improved exchange of experiences at regional level.
- Policy makers should be included in discussions on cover crop use, as these technologies have impacts at various levels and may represent important contributions to the achievement of a more sustainable agriculture in different regions.

Research needs

- technologies for seed production and conservation
- analysis, systematization and documentation of local knowledge on cover crops
- exploration of alternative uses of cover crop grain
- exploration and evaluation of the potential of native species as cover crops
- the positive and negative impacts of cover crops on plant development and yields of perennial crops

- effect of cover crops on pests and diseases in perennial crops
- possible routes for diversification of perennial crops through mechanisms for the recuperation of establishment costs and generation of income in the short, medium and long term
- the use of cover crops for forage and feed for different animal species
- determining the impact of cover crop technologies at community and regional levels.

Conclusions and future strategies

The Merida Workshop conclusions reveal a number of important short-comings in the way cover crops have been promoted up to now in Latin America. Table 27 highlights alternative ways in which different stake-holders may choose to overcome the shortcomings identified. The stake-holders considered here include development interventionists, farmers, extensionists and researchers. Activities considered span the planning, implementation and evaluation phases of a project cycle. They include: developing multi-stakeholder planning, monitoring and evaluation mecha-nisms, taking into account local knowledge in intervention and extension strategies, developing mechanisms and infrastructure for seed distribution, and negotiating among farmers, consumers, or industry over who is to pay for soil and water conservation measures.

One of the major conclusions from the Merida Workshop was that there appears to exist a substantial gap between, on the one side, the rather ideal-istic view of how cover crops might be promoted and the impact they might achieve and, on the other side, the reality encountered by projects engaged in cover crop work. The Workshop conclusions stress the lack of holistic analysis of cover crop impacts (evaluations limited to the plot, and occa-sionally the farm level), and the neglect of local knowledge, which implies the neglect or ignorance of local purposes. Another aspect highlighted in the conclusions is the lack of information exchange and systematization, which reflects the necessity to increase information flow on local, regional, national and wider levels.

Table 27 Shortcomings, actions and actors

Project phase	Shortcomings identified	Action to overcome shortcomings	Actors to be involved
Planning	Failure to take into account different perspectives	Develop multi-stakeholder planning, monitoring and evaluation mechanisms	Development interventionists, Farmers, Extensionists

	Attributes of local knowledge ignored	Develop combined strategies for management practices, e.g. soil and water conservation	Development interventionists, Farmers, Extensionists, Researchers
Implementation	Inequitable distribution of costs and benefits of soil and water management	Consider and negotiate who pays for soil and water conservation measures	Development interventionists, Farmers, Extensionists,
	Top-down transfer of technology	Take *campesino* innovations into account Exchange knowledge, skills and experience at workshops	Extensionists, Researchers
	Poor seed availability	Develop mechanisms and infrastructure for seed distribution networks	Development interventionists, Farmers, Extensionists, Researchers
Outputs and Impact	Full set of cover crop impacts not understood	Systematic evaluation of economic and environmental impact (wider than plot level) of cover crop inclusion and cover crop/livestock interactions	Researchers

We have formulated both functional and purpose-oriented definitions of cover crops. These definitions are derived in concert with the outcomes of an international workshop where different stakeholders were represented. Cover crops have been shown to have different functions, and to be involved in varied biophysical processes and interactions with components of their environment. These functions can be used by farmers for various purposes in their farming system. Aggregating the impact of using cover crops on individual farms over a wide area can produce environmental effects beneficial not only to farmers but also to other people in the same region.

Many different types of cover crops exist, each with specific attributes. Cover crops are flexible in that they can be used for multiple purposes and many are adaptable to different conditions. However, cover crops are not a panacea; they have to be combined with other strategies, and their use is not cost-free.

The promotion and uptake of cover crops has to take into account that they are components of farming systems, that is ecosystems managed for agricultural purposes by people. For the potential of cover crops to be realized, the functions of cover crops have to be understood, the purposes to which they may be put require analysis and planning from a multi-stakeholder perspective, and the evaluation of the impact achieved by incorporating cover crops into agroecosystems should be carried out on multiple dimensions to attain a holistic interpretation.

Papers available on case studies used

Las leguminosas de grano en sistemas de cultivos anuales: caso Honduras. Raúl Aleman, Norman Sagastume, Roberto Zepeda, CIDICCO, Honduras

Investigación y transferencia de tecnología en cultivos de cobertura con cultivos perennes en el CIAT. Benjamin Carreño, Katrin Linzer, Mirtha Orellana, Kate Warren, CIAT, Bolivia

Crianza de cerdos basada en la alimentación no convencional en la reserva de la biosfera de Calakmul, Campeche. Armando Sastre, Aurelio Lopez, Miguel Cervantes, Pronatura Peninsula Yucatán, México

Experiencia de Campesino a Campesino en la zona de amotiguamiento de la reserva de Bosawas. Abelardo Rivas, Jorge Irán; Programa Campesino a Campesino/Union Nacional de Agricultores y Ganaderos, Nicaragua

La Garrotilla (*Medicago hispidia*), una alternativa campesina y agroecologica en el agroecosistema de Cabecera de Valle en Bolivia. Elvira Serano, AGRUCO, Bolivia

List of workshop participants

Bolivia

Elvira Serrano, AGRUCO, Programa Agroecología, Universidad
Cochabamba, Casilla 1280, Cochabamba, Tel (591) 4 52601,
Fax (591) 4 52602, e-mail agruco@dicyt.nrc.edu.bo

Katrin Linzer, Kate Warren, Mirtha Orellana, Benjamin Carreño,
CIAT, Centro de Investigaciones Agroepecuarios Tropicales, Casilla 247,
Santa Cruz de la Sierra, Tel (591) 3 342996, Fax (591) 3 342996,
e-mail ciat@mitai.mrs.bolnet.bo

El Salvador

Carlos Cisneros, Ing. Othmaro Alvarado, Ing. Mario Samayoa,
CENTA, Apdo 2454, San Salvador, Tel (503) 338 4503,
Fax (503) 338 4278 or 298 1670, e-mail agrisost@es.com.sv

EE.UU

Robert Walle, Tel/fax (352) 373 7505,
e-mail afn29600@afn.org

Guatemala

Rafael Ceballos Solares, Proyecto PAS-MAGA/GTZ, Sayatche, Petén,
Tel/fax (502) 9286116

Honduras

Gabino Lopez, COSECHA, Apdo. 3586, Tegucigalpa, Tel/fax (504) 762354

Mario Ardón Mejía, Apdo. 1749, Tegucigalpa, Fax (504) 371291,
e-mail ardon%ceiba@sdnhon.org.hn

Jacqueline Chenier, FUNDACION MISEREOR, Apdo. 1787, Tegucigalpa,
Tel (504) 372719, Fax (504) 371364, e-mail jackie%misereor@sdn.org.hn

Edgardo Padilla, Gaspar Alvarado, a/c ESNACIFOR, CONSEFORH, Apdo. 45, Carretera Norte Siguatepeque, Tel/fax (504) 732770, e-mail conseforh@globalnet.hn

Raúl Alemán, Norman Sagastume, Roberto Zepeda, CIDICCO, Apdo. 4443, Tegucigalpa, Tel/fax (504) 323850, e-mail cidicco@ns.gbm.hn

Pedro Torres, Anastacio Mendez, PROCONDEMA, Apdo. 40, Choluteca, Tel/fax (504) 521123

Pedro Rivera, TOCOA, Casa Rural, Tocoa, Dpto. de Colón , Tel/fax (504) 443912

Manfred Furst, Ernesto Chinchilla, FUNBAHNCAFE/OFICINA DED, Proyecto Merendón, Apdo. 2882, 28 Ave. 9 Calle S.O. #97, San Pedro Sula, Tel/fax (504) 521123

Jon Hellin, Natural Resources Institute (NRI), Apdo. postal 791, Tegucigalpa, Tel (504) 225248, Fax (504) 379628, e-mail hellin@ns.gbm.hn

México

Armando Sastre, PRONATURA PENINSULA YUCATAN, Calle 1 D #254, Mérida 97120, Yucatán, Tel (99) 442290, Fax (99) 443580, e-mail PPY@pibil.finred.com.mx

Daniel González Cortés, Tereso Saldivia, PLAN PILOTO FORESTAL/ ACUERDO MEXICO-ALEMAN, Apdo. 43-B, Chetumal, Q.Roo., Tel (983) 24424, Fax (983) 29269, e-mail ppfi@balam.cuc.uqroo.mx

Manuel Huz, Sergio Zapata Dominguez, PRONATURA CHIAPAS, Av. Benito Juárez No. 11-B, Apdo. 219, San Cristóbal de las Casas 29200, Tel/fax (967) 85000, e-mail 74052.2140@compuserve.com ó pronaturach@laneta.apc.org

Arisbe Mendoza, Francisco Bautista, PROTROPICO, FMVZ-UADY, Apdo. Postal 28 Cordemex, Mérida, Yucatán, Tel/fax (99) 460332

Rafael Chávez, Mario Bernardino, CETAMEX, Apdo. 17, Nochixtlán, Oaxaca , Tel/fax (952) 20009

Victor Villaluazo, Pedro Figueroa, Proyecto Manatlán, Apdo. 64, Autlán, Jalisco 48900, Tel/fax (338) 11165

Ramiro Hernandez, Felipe Tomas, MADERAS DEL PUEBLO DEL SURESTE A.C., Avenida del Norte 1238, Despacho 1, Col. Navarte, Mexico D.F., Tel (5) 6055242, Fax (5) 6055281 or Zanatepec, Oaxaca, Tel/fax (972) 10163 or Matias Romero, Tel/fax (972) 21674

Bernard Triomphe, % CIMMYT, Texcoco, Estado de Mexico, Mexico, e-mail: b.triomphe@cgnet.com

Nicaragua
Abelardo Rivas, Jorge Irán Vasquez, PCaC/UNAG, Apdo. 4526, Managua, Fax (505) 2284028

Paraguay
Nestor Fabian Delgadillo Diaz, Policarpo Gonzalez Garcete, ALA 90, Programa de Colonización Agraria San Pedro y Caaguazu, Defensore del Chaco 237-238-239, Coronel Oviedo - Rca. del Paraguay, Casilla Postal 030, Fax/tel (595) 521 203685, (595) 521 203128, e-mail pdriel@infonet.com.py

United Kingdom
Georg Cadisch, WYE COLLEGE, Wye, Ashford, Kent TN25 5AH, UK, Tel (1233) 812401, Fax (1233) 812855, e-mail g.cadisch@wye.ac.uk

Rob Paterson, NRI, Chatham Maritime, Kent ME4 4TB, UK, Tel (44) 1634 883737, Fax (44) 1634 883888, e-mail r.paterson@gre.ac.uk

Organizers
Simon Anderson, Yucatán:London Inter-University Collaborative Project, FMVZ-UADY, APDO 116-4, Mérida 97100, Yucatán, México, Tel/fax (52) 99-273852, e-mail SiAnderson@compuserve.com
and Wye College, Nr Ashford, Kent TN25 5AH, UK, Tel (44) 1233 812401 , Fax (44) 1233 812855, e-mail Simon.Anderson@ic.ac.uk

Sabine Gündel, Natural Resources Institute, Natural Resource Management Department, Chatham Maritime, Kent ME4 4TB, UK, Tel (44) 1634 883639, Fax (44) 1634 883888, e-mail: s.gundel@gre.ac.uk

Bernadette Keane, FMVZ-UADY, APDO 116-4, Mérida 97100, Yucatán México, Tel/fax (52) 99-273852, dip98@diario1.sureste.com

Barry Pound, NRI, Chatham Maritime, Kent ME4 4TB, UK, Fax (44) 1634 880066, e-mail b.pound@greenwich.ac.uk

List and brief description of cover crop species mentioned*

* (most of the information for this list is sourced from Kiff, Pound and Holdsworth, 1996 and from Binder, 1997).

Arachis pintoi: commonly known as Forage Groundnut; is a tropical perennial legume frequently intercropped with pasture (e.g. *Brachiaria* spp.) or perennial crops or used for land rehabilitation. Shade tolerant. Initially slow to establish. Potential as animal forage. Its origin is Central Brazil but it is distributed in North, Central and South America, as well as Asia and Australia.

Cajanus cajan: an annual, biennial or perennial plant for the humid or semi-arid tropics, depending on varietal selection. Known as Pigeon Pea. Origin India. It is used for human food as well as for animal feed. Other uses are fuelwood, erosion control, and fencing. Once established it is tolerant to dry spells.

Calopogonium caeruleum: a climbing, perennial legume, which is mainly used as cover crop (e.g. under rubber) and green manure. It has a slow establishment but is very persistent. It is more drought tolerant than *Calopogonium mucunoides*.

Calopogonium mucunoides: origin South America. It is used as animal feed and for erosion control. This perennial legume has a vigorous early growth and is good for weed control in plantation crops. It can combat invasive weeds such as *Cyperus* spp. Grown for early cover in pasture mixtures with *Pueraria* and *Centrosema*, which become respectively more dominant with time. Tolerant of high aluminium and manganese. Poor palatability to livestock.

Canavalia ensiformis: an annual, erect growing legume, which originates from Central America and the West Indies. The common name is Jack Bean or Horse Bean. It is consumed by humans and animals (but note toxicity problems). Hardy, versatile drought-resistant plant; intercropped with perennials and annuals.

Centrosema pubescens: a perennial, climbing legume from tropical South America, which adapts to a wide range of soil types. It covers the ground quickly, is fairly resistant to flooding and drought and is shade tolerant. It is consumed by livestock and used as a forage (best grazed in rotation) and green manure, and for erosion control.

Crotalaria juncea: commonly known as Sunn Hemp, an erect, annual crop from India, which is also used for fibre production. It is intercropped with sugarcane, pineapple and coffee as a cover crop, and used to reclaim *Imperata cylindrica*-infested land in east Java. Dried forage is used for livestock or is ploughed under as a green manure.

Desmodium ovalifolium (D. ascendens): a perennial legume from Asia, which is intercropped as a cover crop with plantation crops. Widespread throughout the tropics. It adapts to low-fertility soils and it tolerates shade and heavy grazing. It is used as pasture and as a human medicine.

Flemingia macrophylla: erect perennial shrub. Good weed control by using leaves as mulch (leaves decompose slowly). Low digestibility, but some value as dry season browse. Can be contour planted for erosion control.

Lablab purpureus (Dolichos lab-lab): is a short-lived climbing perennial, usually treated as an annual. Grown mainly for green pods and for livestock fodder (note that dry grain contains cyanogenic glucosides), but also a useful green manure and cover crop. Adapts to a wide range of soils.

Lupinus mutabilis: an erect, annual legume often known as Tarwi, which is used for human consumption and for livestock feed (seeds must be soaked in water for several days before use). It is frost and drought resistant and performs well from 1800 m to 4000 m.

Medicago hispida: a perennial legume, well adapted to tropical highlands. Known as Garrotilla in Bolivia, where it is intercropped with staple crops including potatoes and cereals. It is used for animal consumption and green manure. It is not drought tolerant.

Mucuna pruriens: common names include velvet bean, Bengal bean, terciopelo and fertilizer bean, and includes *M. pruriens* var. *deeringiana* and *M. pruriens* var. *utilis*. This classification has replaced the four previous species *Stizolobium deeringianum, S. aterrinum, S. pruriens* and *S. niveum*. It is a vigorous and versatile annual, climbing legume, which is used principally as a cover crop, but also in some areas as a human food and/or animal feed. It grows well on a wide range of soils and provides large amounts of organic matter.

Neonotonia wightii: commonly known as Glycine. A perennial legume, which is used principally as a pasture legume, but can also be a useful cover crop in perennial plantations. It is frost and drought tolerant and it has an ability to suppress weeds, including *Imperata cylindrica*. Prefers fertile, well-drained soils, although it will tolerate some salinity.

Paspalum conjugatum: a perennial or annual stoloniferous grass used for cover in coconut plantations. Used for grazing only when young.

Pachyrhizus erosus: common name is Yam Bean or Jicama. It is an aggressive perennial legume, which builds a tuber, used for human and livestock consumption. Mature seeds and leaves contain the poison rotenone – used as fish poison and insecticide.

Phaseolus coccineus: annual, climbing legume known as Chinapopo, and intercropped as a cover and food crop with maize in the highlands of Honduras, Mexico and parts of South America. Elsewhere known as scarlet runner bean, and grown as a food pulse.

Phaseolus lunatus: origin Central America and known as Lima bean. It is an annual small-seeded legume, which is used for human food (some varieties contain glucosides) and for fodder. It is a good intercrop, especially with maize. It tends to perform poorly under 20°C.

Phaseolus vulgaris: known as French bean or runner bean. Origin Central America, where many varieties can be found. It is an annual legume, which is mainly used for human food production in temperate and tropical climates. Residues are used for livestock feeding. Grows well as an intercrop.

Pisum sativum: known as Field Pea, is an annual, climbing legume distributed in temperate, tropical and subtropical regions. It is used as human food and residues provide livestock feed.

Psophocarpus tetragonolobus: Wing bean. Although a perennial, it is usually grown as an annual as a sole crop, or as intercrop with cereals. Immature pods are harvested as a vegetable. Residues are used as livestock fodder. Nodulates very well. It grows in the humid tropics.

Pueraria phaseoloides: known as Tropical Kudzu. A climbing, perennial legume, which performs best with rainfall over 2500 mm. One of the best tropical legumes for tolerance of waterlogging. Establishes relatively slowly, but then becomes very aggressive, and can become a weed. It is used as pasture and forage for livestock, frequently intercropped with grasses. Can be used for pasture renovation. Used as a cover crop in plantations in the humid tropics.

Raphanus sativus: fodder radish, a non-legume cover crop with rapid establishment for temperate areas, also used as a salad vegetable.

Sesbania bispinosa: an erect, annual legume, grown in sub-humid and humid tropics and sub-tropics. Tolerant of water. Young foliage is used as livestock forage. Other uses include green manure (especially in rice seedbeds), fuel, fibre and erosion control.

Setaria sphacelata: Golden Timothy grass, used for grazing, hay, silage and for ploughing under as a green manure.

Sinapis alba: White mustard. A non-legume, annual cover or green manure crop for temperate conditions. Also used for human food.

Stylosanthes hamata: a legume, which performs as an annual on wetter sites, and perennial on drier sites. Adapted to acidic, infertile soils. It is persistent and resistant to grazing. Used as forage and pasture.

Tephrosia candida: a perennial legume used as a cover crop in lowland plantations to suppress weeds, add fertility and control erosion. Not suitable for human or animal consumption (toxins).

Trifolium repens: a perennial clover, which is used as animal feed and green manure in temperate regions. It tolerates frequent cutting and low rainfall.

Vigna unguiculata: common name is cowpea. It is an erect/climbing annual legume, which is non-aggressive. It is a good intercrop with cereals. The seeds, and sometimes the young leaves, are used for human consumption. The residual hay is used for livestock feed. There is a wide diversity of growth form and period that have potential exploitation for dual-purpose food/cover crops.

Voandzeia subterranea: Bambara groundnut. Annual, creeping legume from West Africa. Will grow on poor soils in hot climates, and is tolerant of dry conditions.

References and bibliography

AGRUCO (1997) El empleo de la garrotilla (Medicago hispida): una tecnologia campesina en el sistema integral de producción agroecologica, Universidad de San Simon, Cochabamba, Bolivia, Poster Presentation at the 'Taller Regional Latino-Americano; Cultivos de Cobertura: Componentes de Sistemas Integrados'; Facultad de Medicina Veterinaria y Zootecnia, Universidad Autonoma de Yucatan, Merida, Mexico, February 3–6, 1997

Alamgir, M. and Arora, P. (1991) Providing food security for all; *IFAD Studies in Rural Poverty No. 1*, Intermediate Technology Publications, London

Altieri, M.A. (1991) Traditional Farming in Latin America, *The Ecologist*, Vol. 21, No. 2 (March/April), pp. 93–96

Anderson, S., Gündel, S., Keane, B. and Pound, B. (1997) Memorias del Taller Regional Latino-Americano; Cultivos de Cobertura: Componentes de Sistemas Integrados; Facultad de Medicina Veterinaria y Zootecnia, Universidad Autonoma de Yucatan, Merida, Mexico, February 3–6

Ardaya, D., Carreño, B., Encinas, C., Huayhua, J.R., Orellana, M. and Warren, K. (1998) 'Manejo de cítricos y piña con cultivos de cobertura'. CIAT, Casilla 247, Santa Cruz, Bolivia

Avila, N.R. and Lopez, J.L. (1990) Sondeo preliminar en la asociacion maiz frijol abono (mucuna spp.) en el Litoral Atlantico de Honduras. Trabajo presentado en la XXXVI Reunion anual del Programa Cooperativo Centroamericano para el Mejoramiento de Cultivos y Animales (PCCMCA), San Salvador, El Salvador

Barber, R. and Navarro, F. (1994) Evaluation of the characteristics of 14 cover crops used in a soil rehabilitation trial; *Land Degradation and Rehabilitation*, Vol. 5, pp. 201–214

Barreto, H.J. (1991) *Evaluation and utilization of different mulches and cover crops for maize production in Central America*, CIMMYT, Guatemala

Barrett, G., Rodenhouse, N. and Bohlen, P. (1990) Role of sustainable agriculture in rural landscapes. In: Edwards, C., Lal, R., Madden, P., Miller, R. and House, G. (eds) *Sustainable agricultural systems*, Ankeny, IA: Soil and Water Conservation Society

Bayer, W. and Waters-Bayer, A. (1998) *Forage Husbandry*, Macmillan Education Ltd, Basingstoke, UK

Belmar, R. and Morris, T.R. (1994) Effects of raw and treated jack bean (Canavalia ensiformis) and of canavanine on the short term feed intake of chicks and pigs. *J.Agric.Sci.Camb.* 123, pp. 407–414

Binder, U. (1997). *Manual de Leguminosas de Nicaragua* (Tomo I y Tomo II). PASOLAC, E.A.G.E, Esteli, Nicaragua

Borget, M. (1992) Food Legumes; *The Tropical Agriculturalist*, CTA, Macmillan, UK

Buckles, D., Salgado, J., Bojoque, H., Antunez, H., Mejia, L., Nolazco, H., de Ramos, L., Medina, G. and Matute, R. (1991) Resultados de la encuesta exploratoria sobre el uso de frijol de abono (Stizolobium deeringianum) en las laderas del litoral de Honduras. In: *Analálisis de los Ensayos Regionales de Agronomía*, 1990, CIMMYT Central American Regional Program Publication, June 1991

Buckles, D. (1994) Velvetbean: A 'new' plant with a history. CIMMYT Internal Document, CIMMYT Economics Program, Mexico

Buckles, D., Triomphe, B. and Sain, G. (1998) *Cover crops in Hillside Agriculture: Farmer Innovation with Mucuna*; International Development Research Centre, Ottawa/International Center for Maize and Wheat Improvement

Bunch, R. and Lopez, G. (1994) Soil recuperation in Central America: measuring the impact four to forty years after intervention; presented at the International Institute for Environment and Development's International Policy Workshop, Bangalore, India

Bunch, R. (1997) Achieving sustainability in the use of green manure. In: *ILEIA Newsletter for Low External Input and Sustainable Agriculture*, Vol. 13, No. 3

Caligari, A. (1995) Leguminosas para adubacao verde de verao no Parana. IAPAR Circular No 80. Londrina : IAPAR, Brasil

Chenier, J. (1997) Especies leguminosas tradicionales como potenciales cultivos de cobertura en el sur de Honduras. Poster Presentation at the

'Taller Regional Latino-Americano; Cultivos de Cobertura: Componentes de Sistemas Integrados'; Facultad de Medicina Veterinaria y Zootecnia, Universidad Autonoma de Yucatan, Merida, Mexico, February 3–6

CIAT/NRI (1997) Informe de actividades del Proyecto 'Investigación Adaptativa en Ichilo-Sara': *Gestión Agrícola* 1996/7. CIAT, Casilla 247, Santa Cruz, Bolivia

CIDICCO, El uso del frijol de abono (Mucuna spp.) como cultivo de cobertura en plantaciones de citricos; *Informe Tecnico*, No. 7, Marzo 1992

CIDICCO, La utilización de leguminosas de cobertura en plantaciones perennes. (Basado en las experiencias de la plantación de palma africana en San Alejo). *Noticias sobre cultivos de cobertura* No. 7, Febrero 1994

CIDICCO (1997), Case study presented at the 'Taller Regional Latino-Americano; Cultivos de Cobertura: Componentes de Sistemas Integrados' by Alemán, R., Paredes, M. and Sagastume, N., Facultad de Medicina Veterinaria y Zootecnia, Universidad Autonoma de Yucatan, Merida, Mexico, February 3–6

CONSEFORH (1997) El Papel de los cultivos de cobertura en plantaciones de investigacion forestal, Poster presentation at the 'Taller Regional Latino-Americano; Cultivos de Cobertura: Componentes de Sistemas Integrados' by Alvarado, G. and Padilla, E., Facultad de Medicina Veterinaria y Zootecnia, Universidad Autonoma de Yucatan, Merida, Mexico, February 3–6

Conway, G. (1997) *The doubly green revolution: Food for all in the 21st century*, Penguin Books, London

Cooper, D., Vellve, R. and Hobbelink, H. (1992) *Growing Diversity: Genetic resources and local food security*, Intermediate Technology Publications, London

Davidson, B.R. and Davidson, H.F. (1993) *Legumes: The Australian experience – the botany, ecology and agriculture of indigenous and immigrant legumes*, Research Studies Press

Decker, A.M., Clark, A.J., Meisinger, J.J., Mulford, F.R. and McIntosh, M.S. (1994) Legume cover crop contribution to no-tillage corn production, *Agronomy Journal*, Vol. 86

Engels, J.M.M. (1995) In situ conservation and sustainable use of plant genetic resources for food and agriculture in developing countries, Report of a DSE/ ATSAF/ IPGRI workshop, May 1995, Bonn, Germany

Farrington, J. (1995) The changing public role in agricultural extension. In: *Food Policy*, Vol. 20, No. 6, Great Britain

Fischler, M. (1996) Research on green manures in Uganda: results from experiments conducted in 1995, Institute of Plant Science, ETH Zurich, Switzerland, Report submitted to the Rockefeller Foundation, Nairobi

Flores, M. (1993) La utilizacion de leguminosas de cobertura en sistemas agricolas tradicionales de Centroamerica; *Informe Tecniico No.5*, CIDICCO, Honduras

Flores, M. (1994) The use of leguminous cover crops in traditional farming systems in Central America. In: Thurston, D. et al. (1994) *Slash/mulch: how farmers us it and what researchers know about it*; Cornell International Institute for Food, Agriculture and Development, Cornell University, New York

Folorunso, O.A., Rolston, D.E., Prichard, T. and Louie, D.T. (1992) Cover crops lower soil surface strength, may improve soil permeability, *California Agriculture*, Vol. 46, No. 6, USA

Francis, D. (1994) *Family agriculture – tradition and transformation*, Earthscan Publications, London

Fürst, M. (1997) El desarrollo sostenible de la zona de reserva del Merendon, Honduras (FUNBANCAFE) Poster presentation at the 'Taller Regional Latino-Americano; Cultivos de Cobertura: Componentes de Sistemas Integrados'; Facultad de Medicina Veterinaria y Zootecnia, Universidad Autonoma de Yucatan, Merida, Mexico, February 3–6

Gold, C.S. and Wightman, J.A. (1991) Effects of Intercropping Groundnut with Sunnhemp on Termite Incidence and Damage in India, *Insect Science and its Application*, Vol. 12, pp. 177–182

Gündel, S. (1998) *Participatory Innovation Development and Diffusion: Adoption and adaptation of introduced legumes in the traditional slash-and-burn peasant farming system in Yucatan, Mexico*. Kommunikation und Beratung. Sozialwissenschaftliche Schriften zur Landnutzung und Entwicklung. Margraf Verlag, No. 21

Gündel, S. (1999) *Participatory Innovation Development and Diffusion: Adoption and adaptation of introduced legumes in the traditional slash-and-burn peasant farming system in Yucatan, Mexico*. TOEB, GTZ, Germany

Gurtino, B., Sitompul, S.M. and van de Heide, J. Reclamation of Alang-Alang land using cover crops on an utisol in Lampung, *AGRIVITA*, Vol. 15. No. 1

Hardy, B. (1993) Cover legumes for Savannas and degraded environments, *CIAT International*, Vol. 12, No.1

Haroon, S.A. and Abadir, S.H. (1989) The Effect of Four Summer Legume Cover Crops on the Population Level of *Meloidogyne inconita, Pratylenchus penetrans* and *Trichodorus christiei*

Holt-Gimenéz, E. (1996) The campesino-a-campesino movement: farmer-led agricultural extension. In: *ODI Agricultural Research and Extension Network*, Network Paper No 59a, January

Hopkinson, D. (1969) Leguminous cover crops for maintaining soil fertility in sisal in Tanzania: Effects on growth and yields, *Expl. Agric.* Vol. 5, Great Britain

Hulugalle, N.R., Lal, R. and Ter Kuile, C.H.H. (1986) Ameloriation of soil physical properties by Mucuna after mechanized land clearing of a tropical forest, *Soil Science*, Vol. 141, No.3, USA

ISCO (1996) Dare-to-Share Fair at the 9th Conference of the International Soil Conservation Organisation, Bonn, 26–30 August 1996, Margraf Verlag

Iwanaga, M. (1995) IPGRI strategies for in situ conservation of agricultural biodiversity. In: Engels (1995) In situ conservation and sustainable use of plant genetic resources for food and agriculture in developing countries; Report of a DSE/ATSAF/IPGRI workshop, 2–4 May 1995, Bonn, Germany

Keatinge, J.D.H., Aiming Qi, Wheeler, T.R., Ellis, R.H., Craufurd, P.Q. and Summerfield, R.J. (1996) Photothermal effects on the phenology of annual legume crops with potential for use as cover crops and green manure in tropical and sub-tropical hillside environments. *Field Crop Abstracts* 49, pp. 1119–1130

Keatinge, J.D.H., Wheeler, T.R., Shah, P.B., Subedi, M., Musitwa, F., Cespedes, E., Aiming Qi, Ellis, R.H. and Summerfield, R.J. (1997) Potential for and constraints to the production of multi-purpose cover crop legumes in hillside environments in key Department For International Development (DFID) target countries; *DFID Project R 6447 Appraisal Report*, Department of Agriculture, The University of Reading, 46 pp

Kessler, C.D.J. (1990) An agronomic evaluation of Jackbean (Canavalia ensiformis) in Yucatan, Mexico; *Expl. Agric.*, Vol. 26, p. 11–22

Kiff, L., Pound, B. and Holdsworth, R. (1996) *Cover crops: A review and database for field users*. Natural Resources Institute, Chatham UK

Lal, Rattan (1990) *Soil erosion in the tropics – principles and management*, Department of Agronomy, The Ohio State University, McGraw-Hill, Inc., USA

Linzer, K., Orellana, M., Carreño, B. and Warren, K. (1997) Investigación y Transferencia de Tecnología en Cultivos de Cobertura en Sistemas de Cultivos Perennes. Paper presented at the Latin American Regional Workshop 'Cultivos de Cobertura: componentes de sistemas integrados'. Mérida, Mexico 3–6 February 1997.

Masson, P.H., Goby, J.P. and Anthelme, B. (1990) Potential of subterranean clover pastures sown on cleared shrubland in the French Mediterranean Pyrenees; 6th FAO meeting on Mediterranean pastures and fodder crops, Bari, Italy

McDonagh, J.F., Toomsan, B., Limpinuntana, V. and Giller, K.E. (1995) Grain legumes and green manures as pre-rice crops in Northern Thailand, *Plant and Soil*, Vol. 177, The Netherlands

Monegat, C. (1991) *Plantas de cobertura del suelo: Caracteristicas y manejo en pequenas propiedades*. CIDICCO, Aptdo Postal 4443, Tegucigalpa MDC, Honduras, C.A.

Muchagata, M. G. (1997) The role of Forest Production in Frontier Farming Systems in Eastern Amazonia. University of East Anglia Overseas Development Group. *DEV Occasional Paper OP36*. ISBN 1 898285 713

Nwokolo, S. and Smartt, J. (eds) (1996) *Food and feed from legumes and oilseeds*, Chapman and Hall, London.

Okali, C., Sumberg, J. and Farrington, J. (1994) *Farmer Participatory Research: Rhetoric and Reality*. Intermediate Technology Publications, London

Peters, W. J. and Neuenschwander, L.F. (1988) *Slash and Burn: Farming in the Third World Forest*. University of Idaho Press, Moscow, Idaho

Power, J.F. (1991) Growth characteristics of legume cover crops in a semiarid environment, *Soil Science Soc. Am.* Vol. 55, USA

Pretty, J. (1995) *Regenerating Agriculture: Policies and Practice for sustainability and self-reliance*, Earthscan Publications, London

RAFI (1998) *Human Nature: Agricultural Biodiversity and Farm-based Food Security*; Rural Advancement Foundation International

Reijntes, C., Haverkort, B. and Waters-Bayer, A. (1992) *Farming for the future: an introduction to low-external-input and sustainable agriculture*, ILEIA, The Netherlands

Richards, M. (1997) *Missing a Moving Target? Colonist Development on the Amazon Frontier.* ODI Research Study, ODI, London

Richards, P. (1995) Farmers' knowledge and plant genetic resource management (pp. 52–58). In: Engels, J.M.M. *In situ conservation and sustainable use of plant genetic reources for food and agriculture in developing countries*; Report of a DSE/ ATSAF/ IPGRI workshop, May 1995, Bonn, Germany

Rivas, A. and Zamora, E. (1996) Experiencia de Campesino a Campesino en la zona de amortiguamiento de la reserva de Bosawas. PCaC, Union Nacional de Agricultores y Ganadero, Nicaragua

Rizvi, S.J.H and Rizvi, V. (eds) (1992) *Alleopathy: basic and applied aspects*; Chapman & Hall, London

Sinulingga, W., Tjitrosomo, H.S.S., Pawirosoemardjo, S. and Rumawas, F. (1989) Effect of several cover crops on the intensity and viability of *Rigidoporus lignosus* on rubber trees. *Buletin Perkaretan* (Vol. 71): pp. 6–12

Steinfeld, H., de Haan, C., and Blackburn, H. (1997) 'Livestock–Environment Interactions: issues and options' a study sponsored by the EU, FAO, US AID, World Bank

Stobbs, T.H. (1969) The value of Centrosema pubescens for increasing animal production and improving soil fertility in Northern Uganda, *East African Agricultural and Forestry Journal*

Suwardjo, H., Dariah, A. and Barus, A. (1991) Rehabilitation of degraded land in Indonesia. In: Moldenhauer, W.C., Hudson, N.W., Sheng, T.C. and San-Wei, L. (eds) *Development of conservation farming on hillslopes*, Soil and Water Conservation Society, Ankeny, Iowa, USA

Thurston, D. (1994) *Tapado Slash/mulch: how farmers use it and what researchers know about it.* Cornell International Institute for Food, Agriculture and Development, Cornell University

Thurston, D. (1997) *Slash/Mulch Systems: sustainable methods for tropical agriculture*, Intermediate Technology Publications, London

Thurston, D. and Abawi, G. (1997) Effects of Organic Mulches, Soil Amendments, and Cover Crops on Soil-borne Plant Pathogens and their Root Diseases. In: Thurston, D., *Slash/Mulch Systems: sustainable methods for tropical agriculture*; Intermediate Technology Publications, London

Tracy, M.S. and Coe, H.S. (1918) Velvet Beans; *Farmers Bulletin 962*, United States Department of Agriculture (USDA), Washington

Van der Heide, J. and Hairiah, K. (1989) The Role of Green Manures in Rainfed Farming Systems in the Humid Tropics; *ILEIA Newsletter*, Vol.5; No. 2

Veldhuizen, L. van., Waters-Bayer, A., Ramirez, R., Johnson, D. and Thompson, J. (1997) *Farmers' Research in Practice: Lessons from the field.* Intermediate Technology Publications, London

Versteeg, M.V. and Koudokpon, V. (1990) Mucuna helps control imperata in Southern Benin; *WAFSRN Bulletin* No. 7

Wachholtz, R. (1997) Socioeconomía de sistemas agrícolas en transición hacia una agricultura permanente en la Provincia Ichilo. IP/GTZ-PRODISA, Casilla 2768, Santa Cruz, Bolivia

Wade, M.K. and Sanchez, P.A. (1983) Mulching and green manure applications for continuous crop production in the Amazon Basin, *Agronomy Journal*, Vol. 75

Warren, K. M. (1997) Estudios de Adopción y Adaptación de Tecnologías por parte de Agricultores en Ichilo y Sara. CIAT, Casilla 247, Santa Cruz, Bolivia

Wilson, R.T. (1995) *Livestock production systems.* CTA, Macmillan, London.

www.ingramcontent.com/pod-product-compliance
Lightning Source LLC
Jackson TN
JSHW011411130125
77033JS00024B/956